T0283597

Computational Neuroscience

Computational Neuroscience

Edited by
Kieran Robertson

Larsen & Keller
www.larsen-keller.com

Computational Neuroscience
Edited by Kieran Robertson
ISBN: 978-1-63549-073-2 (Hardback)

© 2017 Larsen & Keller

 **Larsen & Keller**

Published by Larsen and Keller Education,
5 Penn Plaza,
19th Floor,
New York, NY 10001, USA

Cataloging-in-Publication Data

Computational neuroscience / edited by Kieran Robertson.
    p. cm.
Includes bibliographical references and index.
ISBN 978-1-63549-073-2
1. Computational neuroscience. 2. Neurosciences. 3. Computational biology.
4. Artificial intelligence. I. Robertson, Kieran.
QP357.5 .C66 2017
612.8--dc23

This book contains information obtained from authentic and highly regarded sources. All chapters are published with permission under the Creative Commons Attribution Share Alike License or equivalent. A wide variety of references are listed. Permissions and sources are indicated; for detailed attributions, please refer to the permissions page. Reasonable efforts have been made to publish reliable data and information, but the authors, editors and publisher cannot assume any responsibility for the vailidity of all materials or the consequences of their use.

Trademark Notice: All trademarks used herein are the property of their respective owners. The use of any trademark in this text does not vest in the author or publisher any trademark ownership rights in such trademarks, nor does the use of such trademarks imply any affiliation with or endorsement of this book by such owners.

The publisher's policy is to use permanent paper from mills that operate a sustainable forestry policy. Furthermore, the publisher ensures that the text paper and cover boards used have met acceptable environmental accreditation standards.

Printed and bound in the United States of America.

For more information regarding Larsen and Keller Education and its products, please visit the publisher's website www.larsen-keller.com

Table of Contents

Preface

As a branch of computational science, theoretical neuroscience or computational neuroscience, deals with the study of the brain with regards to its information processing properties and functions. This field examines the brain while it processes information. Thus, the subject is very important for the study of cognition, learning theory and computational modeling and imaging. This field brings together different areas like psychology, cognitive science, computer science, neuroscience, mathematics, etc. This book outlines the processes and applications of computational neuroscience in detail. The various sub-fields along with technological progress that have future implications are glanced at in it. This textbook attempts to assist those with a goal of delving into the field of computational neuroscience.

To facilitate a deeper understanding of the contents of this book a short introduction of every chapter is written below:

Chapter 1- Computational neuroscience studies brain activities that relate to the processing of information. Information is experienced sensorily and the learning process helps to categorize and retrieve ideas in the thinking process. This chapter introduces the reader to the discipline of computational neuroscience by providing comprehensive information.

Chapter 2- This chapter explores the major approaches and concepts of computational neuroscience that aid in the better understanding of the subject. The topics in the chapter include artificial intelligence, artificial consciousness, neurotechnology, neural oscillation and neural backpropogation. The chapter studies how each of these concepts contribute to computational neurology and the techniques and methods of each concept.

Chapter 3- There are several mathematical models used to explain the electrical activity of neurons. This chapter explores the main theories and models like biological neuron model, efficient coding hypothesis, Hodgkin–Huxley model, binding neuron etc. The chapter provides elaborate and insightful information about each model, its main features, objectives and applications.

Chapter 4- Computational neuroscience is a distinct and complex discipline that uses information technology, neuroscience, psychology, electrical engineering, cognitive science, mathematics and physics. Due to the wide array of subjects that contribute to it, computational neuroscience uses a multitude of interdisciplinary tools and technologies like brain–computer interface, single-unit recording, Bayesian approaches to brain function, neurocomputational speech processing, mind uploading and brain-reading. This chapter provides a comprehensive analysis of the tools and technologies used in this field.

Chapter 5- The research findings of computational neuroscience are utilized in a wide array of allied disciplines like computational anatomy, neuroethology, neural coding, neuroinformatics, neurocybernetics etc. This section explores these fields in-depth and provides a brief background about each. The chapter is a compilation of the various allied branches of computational neurosciences that form an integral part of the broader subject matter.

I owe the completion of this book to the never-ending support of my family, who supported me throughout the project.

Editor

Introduction to Computational Neuroscience

Computational neuroscience studies brain activities that relate to the processing of information. Information is experienced sensorily and the learning process helps to categorize and retrieve ideas in the thinking process. This chapter introduces the reader to the discipline of computational neuroscience by providing comprehensive information.

Computational neuroscience (also theoretical neuroscience) is the study of brain function in terms of the information processing properties of the structures that make up the nervous system. It is an interdisciplinary science that links the diverse fields of neuroscience, cognitive science, and psychology with electrical engineering, computer science, mathematics, and physics.

Computational neuroscience is distinct from psychological connectionism and from learning theories of disciplines such as machine learning, neural networks, and computational learning theory in that it emphasizes descriptions of functional and biologically realistic neurons (and neural systems) and their physiology and dynamics. These models capture the essential features of the biological system at multiple spatial-temporal scales, from membrane currents, proteins, and chemical coupling to network oscillations, columnar and topographic architecture, and learning and memory.

These computational models are used to frame hypotheses that can be directly tested by biological or psychological experiments.

History

The term "computational neuroscience" was introduced by Eric L. Schwartz, who organized a conference, held in 1985 in Carmel, California, at the request of the Systems Development Foundation to provide a summary of the current status of a field which until that point was referred to by a variety of names, such as neural modeling, brain theory and neural networks. The proceedings of this definitional meeting were published in 1990 as the book *Computational Neuroscience*. The first open international meeting focused on Computational Neuroscience was organized by James M. Bower and John Miller in San Francisco, California in 1989 and has continued each year since as the annual CNS meeting The first graduate educational program in computational neuroscience was organized as the Computational and Neural Systems Ph.D. program at the California Institute of Technology in 1985.

The early historical roots of the field can be traced to the work of people such as Louis Lapicque, Hodgkin & Huxley, Hubel & Wiesel, and David Marr, to name a few. Lapicque introduced the integrate and fire model of the neuron in a seminal article published in 1907; this model is still one of the most popular models in computational neuroscience for both cellular and neural networks studies, as well as in mathematical neuroscience because of its simplicity. About 40 years later, Hodgkin & Huxley developed the voltage clamp and created the first biophysical model of the action potential. Hubel & Wiesel discovered that neurons in the primary visual cortex, the first cortical area to process information coming from the retina, have oriented receptive fields and are organized in columns. David Marr's work focused on the interactions between neurons, suggesting computational approaches to the study of how functional groups of neurons within the hippocampus and neocortex interact, store, process, and transmit information. Computational modeling of biophysically realistic neurons and dendrites began with the work of Wilfrid Rall, with the first multicompartmental model using cable theory.

Major Topics

Research in computational neuroscience can be roughly categorized into several lines of inquiry. Most computational neuroscientists collaborate closely with experimentalists in analyzing novel data and synthesizing new models of biological phenomena.

Single-neuron Modeling

Even single neurons have complex biophysical characteristics and can perform computations (e.g.). Hodgkin and Huxley's original model only employed two voltage-sensitive currents (Voltage sensitive ion channels are glycoprotein molecules which extend through the lipid bilayer, allowing ions to traverse under certain conditions through the axolemma), the fast-acting sodium and the inward-rectifying potassium. Though successful in predicting the timing and qualitative features of the action potential, it nevertheless failed to predict a number of important features such as adaptation and shunting. Scientists now believe that there are a wide variety of voltage-sensitive currents, and the implications of the differing dynamics, modulations, and sensitivity of these currents is an important topic of computational neuroscience.

The computational functions of complex dendrites are also under intense investigation. There is a large body of literature regarding how different currents interact with geometric properties of neurons.

Some models are also tracking biochemical pathways at very small scales such as spines or synaptic clefts.

There are many software packages, such as GENESIS and NEURON, that allow rapid and systematic *in silico* modeling of realistic neurons. Blue Brain, a project founded by Henry Markram from the École Polytechnique Fédérale de Lausanne, aims to construct a biophysically detailed simulation of a cortical column on the Blue Gene supercomputer.

A problem in the field is that detailed neuron descriptions are computationally expensive and this can handicap the pursuit of realistic network investigations, where many neurons need to be simulated. So, researchers that study large neural circuits typically represent each neuron and synapse simply, ignoring much of the biological detail. This is unfortunate as there is evidence that the richness of biophysical properties on the single neuron scale can supply mechanisms that serve as the building blocks for network dynamics. Hence there is a drive to produce simplified neuron models that can retain significant biological fidelity at a low computational overhead. Algorithms have been developed to produce faithful, faster running, simplified surrogate neuron models from computationally expensive, detailed neuron models.

Development, Axonal Patterning, and Guidance

How do axons and dendrites form during development? How do axons know where to target and how to reach these targets? How do neurons migrate to the proper position in the central and peripheral systems? How do synapses form? We know from molecular biology that distinct parts of the nervous system release distinct chemical cues, from growth factors to hormones that modulate and influence the growth and development of functional connections between neurons.

Theoretical investigations into the formation and patterning of synaptic connection and morphology are still nascent. One hypothesis that has recently garnered some attention is the *minimal wiring hypothesis*, which postulates that the formation of axons and dendrites effectively minimizes resource allocation while maintaining maximal information storage.

Sensory Processing

Early models of sensory processing understood within a theoretical framework are credited to Horace Barlow. Somewhat similar to the minimal wiring hypothesis described in the preceding section, Barlow understood the processing of the early sensory systems to be a form of efficient coding, where the neurons encoded information which minimized the number of spikes. Experimental and computational work have since supported this hypothesis in one form or another.

Current research in sensory processing is divided among a biophysical modelling of different subsystems and a more theoretical modelling of perception. Current models of perception have suggested that the brain performs some form of Bayesian inference and integration of different sensory information in generating our perception of the physical world.

Memory and Synaptic Plasticity

Earlier models of memory are primarily based on the postulates of Hebbian learning. Biologically relevant models such as Hopfield net have been developed to address the properties of associative, rather than content-addressable, style of memory that occur in biological systems. These attempts are primarily focusing on the formation of medium- and long-term memory, localizing in the hippocampus. Models of working memory, relying on theories of network oscillations and persistent activity, have been built to capture some features of the prefrontal cortex in context-related memory.

One of the major problems in neurophysiological memory is how it is maintained and changed through multiple time scales. Unstable synapses are easy to train but also prone to stochastic disruption. Stable synapses forget less easily, but they are also harder to consolidate. One recent computational hypothesis involves cascades of plasticity that allow synapses to function at multiple time scales. Stereochemically detailed models of the acetylcholine receptor-based synapse with the Monte Carlo method, working at the time scale of microseconds, have been built. It is likely that computational tools will contribute greatly to our understanding of how synapses function and change in relation to external stimulus in the coming decades.

Behaviors of Networks

Biological neurons are connected to each other in a complex, recurrent fashion. These connections are, unlike most artificial neural networks, sparse and usually specific. It is not known how information is transmitted through such sparsely connected networks. It is also unknown what the computational functions of these specific connectivity patterns are, if any.

The interactions of neurons in a small network can be often reduced to simple models such as the Ising model. The statistical mechanics of such simple systems are well-characterized theoretically. There has been some recent evidence that suggests that dynamics of arbitrary neuronal networks can be reduced to pairwise interactions. It is not known, however, whether such descriptive dynamics impart any important computational function. With the emergence of two-photon microscopy and calcium imaging, we now have powerful experimental methods with which to test the new theories regarding neuronal networks.

In some cases the complex interactions between *inhibitory* and *excitatory* neurons can be simplified using mean field theory, which gives rise to the population model of neural networks. While many neurotheorists prefer such models with reduced complexity, others argue that uncovering structural functional relations depends on including as much neuronal and network structure as possible. Models of this type are typically built in large simulation platforms like GENESIS or NEURON. There have been some attempts to provide unified methods that bridge and integrate these levels of complexity.

Cognition, Discrimination, and Learning

Computational modeling of higher cognitive functions has only recently begun. Experimental data comes primarily from single-unit recording in primates. The frontal lobe and parietal lobe function as integrators of information from multiple sensory modalities. There are some tentative ideas regarding how simple mutually inhibitory functional circuits in these areas may carry out biologically relevant computation.

The brain seems to be able to discriminate and adapt particularly well in certain contexts. For instance, human beings seem to have an enormous capacity for memorizing and recognizing faces. One of the key goals of computational neuroscience is to dissect how biological systems carry out these complex computations efficiently and potentially replicate these processes in building intelligent machines.

The brain's large-scale organizational principles are illuminated by many fields, including biology, psychology, and clinical practice. Integrative neuroscience attempts to consolidate these observations through unified descriptive models and databases of behavioral measures and recordings. These are the bases for some quantitative modeling of large-scale brain activity.

The Computational Representational Understanding of Mind (CRUM) is another attempt at modeling human cognition through simulated processes like acquired rule-based systems in decision making and the manipulation of visual representations in decision making.

Consciousness

One of the ultimate goals of psychology/neuroscience is to be able to explain the everyday experience of conscious life. Francis Crick and Christof Koch made some attempts to formulate a consistent framework for future work in neural correlates of consciousness (NCC), though much of the work in this field remains speculative.

Computational Clinical Neuroscience

It is a field that brings together experts in neuroscience, neurology, psychiatry, decision sciences and computational modeling to quantitatively define and investigate problems in neurological and psychiatric diseases, and to train scientists and clinicians that wish to apply these models to diagnosis and treatment.

References

- Bower, James M. (2013). 20 years of Computational neuroscience. Berlin, Germany: Springer. ISBN 978-1461414230.

- Wu, Samuel Miao-sin; Johnston, Daniel (1995). Foundations of cellular neurophysiology. Cambridge, Mass: MIT Press. ISBN 0-262-10053-3.

- Koch, Christof (1999). Biophysics of computation: information processing in single neurons. Oxford [Oxfordshire]: Oxford University Press. ISBN 0-19-510491-9.

- Anderson, Charles H.; Eliasmith, Chris (2004). Neural Engineering: Computation, Representation, and Dynamics in Neurobiological Systems (Computational Neuroscience). Cambridge, Mass: The MIT Press. ISBN 0-262-55060-1.

- Forrest MD (April 2015). "Simulation of alcohol action upon a detailed Purkinje neuron model and a simpler surrogate model that runs >400 times faster". BMC Neuroscience. 16 (27). doi:10.1186/s12868-015-0162-6.

- Friston KJ, Stephan KE, Montague R, Dolan RJ (2014). "Computational psychiatry: the brain as a phantastic organ". Lancet Psych. 1 (2): 148–58. doi:10.1016/S2215-0366(14)70275-5.

Essential Concepts and Approaches of Computational Neuroscience

2

This chapter explores the major approaches and key concepts of computational neuroscience that aid in the better understanding of the subject. The topics in the chapter include artificial intelligence, artificial consciousness, neurotechnology, neural oscillation and neural backpropogation. The chapter studies how each of these concepts contribute to computational neurology and the techniques and methods of each concept.

Artificial Intelligence

Artificial intelligence (AI) is intelligence exhibited by machines. In computer science, an ideal "intelligent" machine is a flexible rational agent that perceives its environment and takes actions that maximize its chance of success at some goal. Colloquially, the term "artificial intelligence" is applied when a machine mimics "cognitive" functions that humans associate with other human minds, such as "learning" and "problem solving". As machines become increasingly capable, facilities once thought to require intelligence are removed from the definition. For example, optical character recognition is no longer perceived as an exemplar of "artificial intelligence" having become a routine technology. Capabilities still classified as AI include advanced Chess and Go systems and self-driving cars.

AI research is divided into subfields that focus on specific problems or on specific approaches or on the use of a particular tool or towards satisfying particular applications.

The central problems (or goals) of AI research include reasoning, knowledge, planning, learning, natural language processing (communication), perception and the ability to move and manipulate objects. General intelligence is among the field's long-term goals. Approaches include statistical methods, computational intelligence, soft computing (e.g. machine learning), and traditional symbolic AI. Many tools are used in AI, including versions of search and mathematical optimization, logic, methods based on probability and economics. The AI field draws upon computer science, mathematics, psychology, linguistics, philosophy, neuroscience and artificial psychology.

The field was founded on the claim that human intelligence "can be so precisely described that a machine can be made to simulate it." This raises philosophical arguments about the nature of the mind and the ethics of creating artificial beings endowed with human-like intelligence, issues which have been explored by myth, fiction and philosophy since antiquity. Attempts to create artificial intelligence has experienced many setbacks, including the ALPAC report of 1966, the abandonment of perceptrons in 1970, the Lighthill Report of 1973 and the collapse of the Lisp machine market in 1987. In the twenty-first century AI techniques became an essential part of the technology industry, helping to solve many challenging problems in computer science.

History

While the concept of thought capable artificial beings appeared as storytelling devices in antiquity, the idea of actually trying to build a machine to perform useful reasoning may have begun with Ramon Lull (c. 1300 CE). With his Calculus ratiocinator, Gottfried Leibniz extended the concept of the calculating machine (Wilhelm Schickard engineered the first one around 1623), intending to perform operations on concepts rather than numbers. Since the 19th century, artificial beings are common in fiction, as in Mary Shelley's *Frankenstein* or Karel Čapek's *R.U.R. (Rossum's Universal Robots)*.

The study of mechanical or "formal" reasoning began with philosophers and mathematicians in antiquity. In the 19th century, George Boole refined those ideas into propositional logic and Gottlob Frege developed a notational system for mechanical reasoning (a *"predicate calculus"*). Around the 1940s, Alan Turing's theory of computation suggested that a machine, by shuffling symbols as simple as "0" and "1", could simulate any conceivable act of mathematical deduction. This insight, that digital computers can simulate any process of formal reasoning, is known as the Church–Turing thesis. Along with concurrent discoveries in neurology, information theory and cybernetics, this led researchers to consider the possibility of building an electronic brain. The first work that is now generally recognized as AI was McCullough and Pitts' 1943 formal design for Turing-complete "artificial neurons".

The field of AI research was founded at a conference at Dartmouth College in 1956. The attendees, including John McCarthy, Marvin Minsky, Allen Newell, Arthur Samuel and Herbert Simon, became the leaders of AI research. They and their students wrote programs that were, to most people, simply astonishing: computers were winning at checkers, solving word problems in algebra, proving logical theorems and speaking English. By the middle of the 1960s, research in the U.S. was heavily funded by the Department of Defense and laboratories had been established around the world. AI's founders were optimistic about the future: Herbert Simon predicted that "machines will be capable, within twenty years, of doing any work a man can do". Marvin Minsky agreed, writing that "within a generation ... the problem of creating 'artificial intelligence' will substantially be solved".

They failed to recognize the difficulty of some of the remaining tasks. Progress slowed and in 1974, in response to the criticism of Sir James Lighthill and ongoing pressure from the US Congress to fund more productive projects, both the U.S. and British governments cut off exploratory research in AI. The next few years would later be called an "AI winter", a period when funding for AI projects was hard to find.

In the early 1980s, AI research was revived by the commercial success of expert systems, a form of AI program that simulated the knowledge and analytical skills of human experts. By 1985 the market for AI had reached over a billion dollars. At the same time, Japan's fifth generation computer project inspired the U.S and British governments to restore funding for academic research. However, beginning with the collapse of the Lisp Machine market in 1987, AI once again fell into disrepute, and a second, longer-lasting hiatus began.

In the late 1990s and early 21st century, AI began to be used for logistics, data mining, medical diagnosis and other areas. The success was due to increasing computational power, greater emphasis on solving specific problems, new ties between AI and other fields and a commitment by researchers to mathematical methods and scientific standards. Deep Blue became the first computer chess-playing system to beat a reigning world chess champion, Garry Kasparov on 11 May 1997.

Advanced statistical techniques (loosely known as deep learning), access to large amounts of data and faster computers enabled advances in machine learning and perception. By the mid 2010s, machine learning applications were used throughout the world. In a *Jeopardy!* quiz show exhibition match, IBM's question answering system, Watson, defeated the two greatest Jeopardy champions, Brad Rutter and Ken Jennings, by a significant margin. The Kinect, which provides a 3D body–motion interface for the Xbox 360 and the Xbox One use algorithms that emerged from lengthy AI research as do intelligent personal assistants in smartphones. In March 2016, AlphaGo won 4 out of 5 games of Go in a match with Go champion Lee Sedol, becoming the first computer Go-playing system to beat a professional Go player without handicaps.

Research

Goals

The general problem of simulating (or creating) intelligence has been broken down into sub-problems. These consist of particular traits or capabilities that researchers expect an intelligent system to display. The traits described below have received the most attention.

Deduction, Reasoning, Problem Solving

Early researchers developed algorithms that imitated step-by-step reasoning that humans use when they solve puzzles or make logical deductions (reason). By the late 1980s and 1990s, AI research had developed methods for dealing with uncertain or incomplete information, employing concepts from probability and economics.

For difficult problems, algorithms can require enormous computational resources—most experience a "combinatorial explosion": the amount of memory or computer time required becomes astronomical for problems of a certain size. The search for more efficient problem-solving algorithms is a high priority.

Human beings ordinarily use fast, intuitive judgments rather than step-by-step deduction that early AI research was able to model. AI has progressed using "sub-symbolic" problem solving: embodied agent approaches emphasize the importance of sensorimotor skills to higher reasoning; neural net research attempts to simulate the structures inside the brain that give rise to this skill; statistical approaches to AI mimic the human ability to guess.

Knowledge Representation

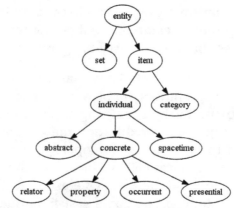

An ontology represents knowledge as a set of concepts within a domain and the relationships between those concepts.

Knowledge representation and knowledge engineering are central to AI research. Many of the problems machines are expected to solve will require extensive knowledge about the world. Among the things that AI needs to represent are: objects, properties, categories and relations between objects; situations, events, states and time; causes and effects; knowledge about knowledge (what we know about what other people know); and many other, less well researched domains. A representation of "what exists" is an ontology: the set of objects, relations, concepts and so on that the machine knows about. The most general are called upper ontologies, which attempt to provide a foundation for all other knowledge.

Among the most difficult problems in knowledge representation are:

Default reasoning and the qualification problem

> Many of the things people know take the form of "working assumptions." For example, if a bird comes up in conversation, people typically picture an animal that is fist sized, sings, and flies. None of these things are true about all birds.

John McCarthy identified this problem in 1969 as the qualification problem: for any commonsense rule that AI researchers care to represent, there tend to be a huge number of exceptions. Almost nothing is simply true or false in the way that abstract logic requires. AI research has explored a number of solutions to this problem.

The breadth of commonsense knowledge

The number of atomic facts that the average person knows is astronomical. Research projects that attempt to build a complete knowledge base of commonsense knowledge (e.g., Cyc) require enormous amounts of laborious ontological engineering—they must be built, by hand, one complicated concept at a time. A major goal is to have the computer understand enough concepts to be able to learn by reading from sources like the Internet, and thus be able to add to its own ontology.

The subsymbolic form of some commonsense knowledge

Much of what people know is not represented as "facts" or "statements" that they could express verbally. For example, a chess master will avoid a particular chess position because it "feels too exposed" or an art critic can take one look at a statue and instantly realize that it is a fake. These are intuitions or tendencies that are represented in the brain non-consciously and sub-symbolically. Knowledge like this informs, supports and provides a context for symbolic, conscious knowledge. As with the related problem of sub-symbolic reasoning, it is hoped that situated AI, computational intelligence, or statistical AI will provide ways to represent this kind of knowledge.

Planning

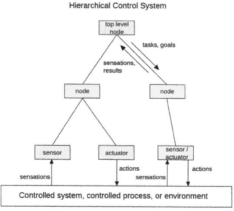

A hierarchical control system is a form of control system in which a set of devices and governing software is arranged in a hierarchy.

Intelligent agents must be able to set goals and achieve them. They need a way to visualize the future (they must have a representation of the state of the world and be able to make predictions about how their actions will change it) and be able to make choices that maximize the utility (or "value") of the available choices.

In classical planning problems, the agent can assume that it is the only thing acting on the world and it can be certain what the consequences of its actions may be. However, if the agent is not the only actor, it must periodically ascertain whether the world matches its predictions and it must change its plan as this becomes necessary, requiring the agent to reason under uncertainty.

Multi-agent planning uses the cooperation and competition of many agents to achieve a given goal. Emergent behavior such as this is used by evolutionary algorithms and swarm intelligence.

Learning

Machine learning is the study of computer algorithms that improve automatically through experience and has been central to AI research since the field's inception.

Unsupervised learning is the ability to find patterns in a stream of input. Supervised learning includes both classification and numerical regression. Classification is used to determine what category something belongs in, after seeing a number of examples of things from several categories. Regression is the attempt to produce a function that describes the relationship between inputs and outputs and predicts how the outputs should change as the inputs change. In reinforcement learning the agent is rewarded for good responses and punished for bad ones. The agent uses this sequence of rewards and punishments to form a strategy for operating in its problem space. These three types of learning can be analyzed in terms of decision theory, using concepts like utility. The mathematical analysis of machine learning algorithms and their performance is a branch of theoretical computer science known as computational learning theory.

Within developmental robotics, developmental learning approaches were elaborated for lifelong cumulative acquisition of repertoires of novel skills by a robot, through autonomous self-exploration and social interaction with human teachers, and using guidance mechanisms such as active learning, maturation, motor synergies, and imitation.

Natural Language Processing (Communication)

Natural language processing gives machines the ability to read and understand the languages that humans speak. A sufficiently powerful natural language processing system would enable natural language user interfaces and the acquisition of knowledge directly from human-written sources, such as newswire texts. Some straightforward applications of natural language processing include information retrieval, text mining, question answering and machine translation.

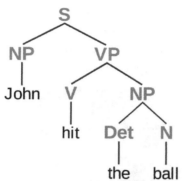

A parse tree represents the syntactic structure of a sentence according to some formal grammar.

A common method of processing and extracting meaning from natural language is through semantic indexing. Increases in processing speeds and the drop in the cost of data storage makes indexing large volumes of abstractions of the user's input much more efficient.

Perception

Machine perception is the ability to use input from sensors (such as cameras, microphones, tactile sensors, sonar and others more exotic) to deduce aspects of the world. Computer vision is the ability to analyze visual input. A few selected subproblems are speech recognition, facial recognition and object recognition.

Motion and Manipulation

The field of robotics is closely related to AI. Intelligence is required for robots to be able to handle such tasks as object manipulation and navigation, with sub-problems of localization (knowing where you are, or finding out where other things are), mapping (learning what is around you, building a map of the environment), and motion planning (figuring out how to get there) or path planning (going from one point in space to another point, which may involve compliant motion – where the robot moves while maintaining physical contact with an object).

Long-term Goals

Among the long-term goals in the research pertaining to artificial intelligence are: (1) Social intelligence, (2) Creativity, and (3) General intelligence.

Social Intelligence

Affective computing is the study and development of systems and devices that can recognize, interpret, process, and simulate human affects. It is an interdisciplinary field spanning computer sciences, psychology, and cognitive science. While the origins of the field may be traced as far back as to early philosophical inquiries into emotion, the more modern branch of computer science originated with Rosalind Picard's 1995 paper on affective computing. A motivation for the research is the ability to simulate empathy. The machine should interpret the emotional state of humans and adapt its behaviour to them, giving an appropriate response for those emotions.

Kismet, a robot with rudimentary social skills

Emotion and social skills play two roles for an intelligent agent. First, it must be able to predict the actions of others, by understanding their motives and emotional states. (This involves elements of game theory, decision theory, as well as the ability to model human emotions and the perceptual skills to detect emotions.) Also, in an effort to facilitate human-computer interaction, an intelligent machine might want to be able to *display* emotions—even if it does not actually experience them itself—in order to appear sensitive to the emotional dynamics of human interaction.

Creativity

A sub-field of AI addresses creativity both theoretically (from a philosophical and psychological perspective) and practically (via specific implementations of systems that generate outputs that can be considered creative, or systems that identify and assess creativity). Related areas of computational research are Artificial intuition and Artificial thinking.

General Intelligence

Many researchers think that their work will eventually be incorporated into a machine with *general* intelligence (known as strong AI), combining all the skills above and exceeding human abilities at most or all of them. A few believe that anthropomorphic features like artificial consciousness or an artificial brain may be required for such a project.

Many of the problems above may require general intelligence to be considered solved. For example, even a straightforward, specific task like machine translation requires that the machine read and write in both languages (NLP), follow the author's argument (reason), know what is being talked about (knowledge), and faithfully reproduce the author's intention (social intelligence). A problem like machine translation is considered "AI-complete". In order to solve this particular problem, one must solve all the problems.

Approaches

There is no established unifying theory or paradigm that guides AI research. Researchers disagree about many issues. A few of the most long standing questions that have remained unanswered are these: should artificial intelligence simulate natural intelligence by studying psychology or neurology? Or is human biology as irrelevant to AI research as bird biology is to aeronautical engineering? Can intelligent behavior be described using simple, elegant principles (such as logic or optimization)? Or does it necessarily require solving a large number of completely unrelated problems? Can intelligence be reproduced using high-level symbols, similar to words and ideas? Or does it require "sub-symbolic" processing? John Haugeland, who coined the term GOFAI (Good Old-Fashioned Artificial Intelligence), also proposed that AI should more properly be referred to as synthetic intelligence, a term which has since been adopted by some non-GOFAI researchers.

Cybernetics and Brain Simulation

In the 1940s and 1950s, a number of researchers explored the connection between neurology, information theory, and cybernetics. Some of them built machines that used electronic networks to exhibit rudimentary intelligence, such as W. Grey Walter's turtles and the Johns Hopkins Beast. Many of these researchers gathered for meetings of the Teleological Society at Princeton University and the Ratio Club in England. By 1960, this approach was largely abandoned, although elements of it would be revived in the 1980s.

Symbolic

When access to digital computers became possible in the middle 1950s, AI research began to explore the possibility that human intelligence could be reduced to symbol manipulation. The research was centered in three institutions: Carnegie Mellon University, Stanford and MIT, and each one developed its own style of research. John Haugeland named these approaches to AI "good old fashioned AI" or "GOFAI". During the 1960s, symbolic approaches had achieved great success at simulating high-level thinking in small demonstration programs. Approaches based on cybernetics or neural networks were abandoned or pushed into the background. Researchers in the 1960s and the 1970s were convinced that symbolic approaches would eventually succeed in creating a machine with artificial general intelligence and considered this the goal of their field.

Cognitive simulation

> Economist Herbert Simon and Allen Newell studied human problem-solving skills and attempted to formalize them, and their work laid the foundations of the field of artificial intelligence, as well as cognitive science, operations re-

search and management science. Their research team used the results of psychological experiments to develop programs that simulated the techniques that people used to solve problems. This tradition, centered at Carnegie Mellon University would eventually culminate in the development of the Soar architecture in the middle 1980s.

Logic-based

Unlike Newell and Simon, John McCarthy felt that machines did not need to simulate human thought, but should instead try to find the essence of abstract reasoning and problem solving, regardless of whether people used the same algorithms. His laboratory at Stanford (SAIL) focused on using formal logic to solve a wide variety of problems, including knowledge representation, planning and learning. Logic was also the focus of the work at the University of Edinburgh and elsewhere in Europe which led to the development of the programming language Prolog and the science of logic programming.

"Anti-logic" or "scruffy"

Researchers at MIT (such as Marvin Minsky and Seymour Papert) found that solving difficult problems in vision and natural language processing required ad-hoc solutions – they argued that there was no simple and general principle (like logic) that would capture all the aspects of intelligent behavior. Roger Schank described their "anti-logic" approaches as "scruffy" (as opposed to the "neat" paradigms at CMU and Stanford). Commonsense knowledge bases (such as Doug Lenat's Cyc) are an example of "scruffy" AI, since they must be built by hand, one complicated concept at a time.

Knowledge-based

When computers with large memories became available around 1970, researchers from all three traditions began to build knowledge into AI applications. This "knowledge revolution" led to the development and deployment of expert systems (introduced by Edward Feigenbaum), the first truly successful form of AI software. The knowledge revolution was also driven by the realization that enormous amounts of knowledge would be required by many simple AI applications.

Sub-symbolic

By the 1980s progress in symbolic AI seemed to stall and many believed that symbolic systems would never be able to imitate all the processes of human cognition, especially perception, robotics, learning and pattern recognition. A number of researchers began to look into "sub-symbolic" approaches to specific AI problems. Sub-symbolic methods manage to approach intelligence without specific representations of knowledge.

Bottom-up, embodied, situated, behavior-based or nouvelle AI

> Researchers from the related field of robotics, such as Rodney Brooks, rejected symbolic AI and focused on the basic engineering problems that would allow robots to move and survive. Their work revived the non-symbolic viewpoint of the early cybernetics researchers of the 1950s and reintroduced the use of control theory in AI. This coincided with the development of the embodied mind thesis in the related field of cognitive science: the idea that aspects of the body (such as movement, perception and visualization) are required for higher intelligence.

Computational intelligence and soft computing

> Interest in neural networks and "connectionism" was revived by David Rumelhart and others in the middle 1980s. Neural networks are an example of soft computing --- they are solutions to problems which cannot be solved with complete logical certainty, and where an approximate solution is often enough. Other soft computing approaches to AI include fuzzy systems, evolutionary computation and many statistical tools. The application of soft computing to AI is studied collectively by the emerging discipline of computational intelligence.

Statistical

In the 1990s, AI researchers developed sophisticated mathematical tools to solve specific subproblems. These tools are truly scientific, in the sense that their results are both measurable and verifiable, and they have been responsible for many of AI's recent successes. The shared mathematical language has also permitted a high level of collaboration with more established fields (like mathematics, economics or operations research). Stuart Russell and Peter Norvig describe this movement as nothing less than a "revolution" and "the victory of the neats." Critics argue that these techniques (with few exceptions) are too focused on particular problems and have failed to address the long-term goal of general intelligence. There is an ongoing debate about the relevance and validity of statistical approaches in AI, exemplified in part by exchanges between Peter Norvig and Noam Chomsky.

Integrating the Approaches

Intelligent agent paradigm

> An intelligent agent is a system that perceives its environment and takes actions which maximize its chances of success. The simplest intelligent agents are programs that solve specific problems. More complicated agents include human beings and organizations of human beings (such as firms). The paradigm gives researchers license to study isolated problems and find solutions that are both verifiable and useful, without agreeing on one single approach. An agent that

solves a specific problem can use any approach that works – some agents are symbolic and logical, some are sub-symbolic neural networks and others may use new approaches. The paradigm also gives researchers a common language to communicate with other fields—such as decision theory and economics—that also use concepts of abstract agents. The intelligent agent paradigm became widely accepted during the 1990s.

Agent architectures and cognitive architectures

Researchers have designed systems to build intelligent systems out of inter-acting intelligent agents in a multi-agent system. A system with both symbolic and sub-symbolic components is a hybrid intelligent system, and the study of such systems is artificial intelligence systems integration. A hierarchical control system provides a bridge between sub-symbolic AI at its lowest, reactive levels and traditional symbolic AI at its highest levels, where relaxed time constraints permit planning and world modelling. Rodney Brooks' subsumption architec-ture was an early proposal for such a hierarchical system.

Tools

In the course of 50 years of research, AI has developed a large number of tools to solve the most difficult problems in computer science. A few of the most general of these methods are discussed below.

Search and Optimization

Many problems in AI can be solved in theory by intelligently searching through many possible solutions: Reasoning can be reduced to performing a search. For example, logical proof can be viewed as searching for a path that leads from premises to con-clusions, where each step is the application of an inference rule. Planning algorithms search through trees of goals and subgoals, attempting to find a path to a target goal, a process called means-ends analysis. Robotics algorithms for moving limbs and grasp-ing objects use local searches in configuration space. Many learning algorithms use search algorithms based on optimization.

Simple exhaustive searches are rarely sufficient for most real world problems: the search space (the number of places to search) quickly grows to astronomical numbers. The result is a search that is too slow or never completes. The solution, for many prob-lems, is to use "heuristics" or "rules of thumb" that eliminate choices that are unlikely to lead to the goal (called "pruning the search tree"). Heuristics supply the program with a "best guess" for the path on which the solution lies. Heuristics limit the search for solutions into a smaller sample size.

A very different kind of search came to prominence in the 1990s, based on the math-ematical theory of optimization. For many problems, it is possible to begin the search with some form of a guess and then refine the guess incrementally until no more refinements can be made. These algorithms can be visualized as blind hill climbing: we begin the search at a random point on the landscape, and then, by jumps or steps, we keep moving our guess uphill, until we reach the top. Other optimization algorithms are simulated annealing, beam search and random optimization.

Evolutionary computation uses a form of optimization search. For example, they may begin with a population of organisms (the guesses) and then allow them to mutate and recombine, selecting only the fittest to survive each generation (refining the guesses). Forms of evolutionary computation include swarm intelligence algorithms (such as ant colony or particle swarm optimization) and evolutionary algorithms (such as genetic algorithms, gene expression programming, and genetic programming).

Logic

Logic is used for knowledge representation and problem solving, but it can be applied to other problems as well. For example, the satplan algorithm uses logic for planning and inductive logic programming is a method for learning.

Several different forms of logic are used in AI research. Propositional or sentential logic is the logic of statements which can be true or false. First-order logic also allows the use of quantifiers and predicates, and can express facts about objects, their properties, and their relations with each other. Fuzzy logic, is a version of first-order logic which allows the truth of a statement to be represented as a value between 0 and 1, rather than simply True (1) or False (0). Fuzzy systems can be used for uncertain reasoning and have been widely used in modern industrial and consumer product control systems. Subjective logic models uncertainty in a different and more explicit manner than fuzzy-logic: a given binomial opinion satisfies belief + disbelief + uncertainty = 1 within a Beta distribution. By this method, ignorance can be distinguished from probabilistic statements that an agent makes with high confidence.

Default logics, non-monotonic logics and circumscription are forms of logic designed to help with default reasoning and the qualification problem. Several extensions of logic have been designed to handle specific domains of knowledge, such as: description logics; situation calculus, event calculus and fluent calculus (for representing events and time); causal calculus; belief calculus; and modal logics.

Probabilistic Methods for Uncertain Reasoning

Many problems in AI (in reasoning, planning, learning, perception and robotics) require the agent to operate with incomplete or uncertain information. AI researchers have devised a number of powerful tools to solve these problems using methods from probability theory and economics.

Bayesian networks are a very general tool that can be used for a large number of problems: reasoning (using the Bayesian inference algorithm), learning (using the expectation-maximization algorithm), planning (using decision networks) and perception (using dynamic Bayesian networks). Probabilistic algorithms can also be used for filtering, prediction, smoothing and finding explanations for streams of data, helping perception systems to analyze processes that occur over time (e.g., hidden Markov models or Kalman filters).

A key concept from the science of economics is "utility": a measure of how valuable something is to an intelligent agent. Precise mathematical tools have been developed that analyze how an agent can make choices and plan, using decision theory, decision analysis, and information value theory. These tools include models such as Markov decision processes, dynamic decision networks, game theory and mechanism design.

Classifiers and Statistical Learning Methods

The simplest AI applications can be divided into two types: classifiers ("if shiny then diamond") and controllers ("if shiny then pick up"). Controllers do, however, also classify conditions before inferring actions, and therefore classification forms a central part of many AI systems. Classifiers are functions that use pattern matching to determine a closest match. They can be tuned according to examples, making them very attractive for use in AI. These examples are known as observations or patterns. In supervised learning, each pattern belongs to a certain predefined class. A class can be seen as a decision that has to be made. All the observations combined with their class labels are known as a data set. When a new observation is received, that observation is classified based on previous experience.

A classifier can be trained in various ways; there are many statistical and machine learning approaches. The most widely used classifiers are the neural network, kernel methods such as the support vector machine, k-nearest neighbor algorithm, Gaussian mixture model, naive Bayes classifier, and decision tree. The performance of these classifiers have been compared over a wide range of tasks. Classifier performance depends greatly on the characteristics of the data to be classified. There is no single classifier that works best on all given problems; this is also referred to as the "no free lunch" theorem. Determining a suitable classifier for a given problem is still more an art than science.

Neural Networks

The study of non-learning artificial neural networks began in the decade before the field of AI research was founded, in the work of Walter Pitts and Warren McCullough. Frank Rosenblatt invented the perceptron, a learning network with a single layer, similar to the old concept of linear regression. Early pioneers also include Alexey Grigorevich Ivakhnenko, Teuvo Kohonen, Stephen Grossberg, Kunihiko Fukushima, Christoph von der Malsburg, David Willshaw, Shun-Ichi Amari, Bernard Widrow, John Hopfield, and others.

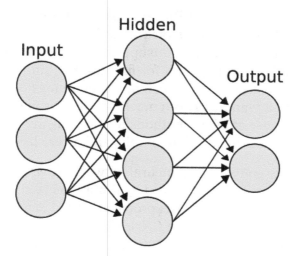

A neural network is an interconnected group of nodes, akin to the vast network of neurons in the human brain.

The main categories of networks are acyclic or feedforward neural networks (where the signal passes in only one direction) and recurrent neural networks (which allow feedback and short-term memories of previous input events). Among the most popular feedforward networks are perceptrons, multi-layer perceptrons and radial basis networks. Neural networks can be applied to the problem of intelligent control (for robotics) or learning, using such techniques as Hebbian learning, GMDH or competitive learning.

Today, neural networks are often trained by the backpropagation algorithm, which had been around since 1970 as the reverse mode of automatic differentiation published by Seppo Linnainmaa, and was introduced to neural networks by Paul Werbos.

Hierarchical temporal memory is an approach that models some of the structural and algorithmic properties of the neocortex.

Deep Feedforward Neural Networks

Deep learning in artificial neural networks with many layers has transformed many important subfields of artificial intelligence, including computer vision, speech recognition, natural language processing and others.

According to a survey, the expression "Deep Learning" was introduced to the Machine Learning community by Rina Dechter in 1986 and gained traction after Igor Aizenberg and colleagues introduced it to Artificial Neural Networks in 2000. The first functional Deep Learning networks were published by Alexey Grigorevich Ivakhnenko and V. G. Lapa in 1965. These networks are trained one layer at a time. Ivakhnenko's 1971 paper describes the learning of a deep feedforward multilayer perceptron with eight layers, already much deeper than many later networks. In 2006, a publication by Geoffrey Hinton and Ruslan Salakhutdinov introduced another way of pre-training many-layered feedforward neural networks (FNNs) one layer at a time, treating each layer in turn as an unsupervised restricted Boltzmann machine, then using supervised backpropagation for fine-tuning. Similar to shallow artificial neural networks, deep neural networks can model complex non-linear relationships. Over the last few years, advances in both machine learning algorithms and computer hardware have led to more efficient methods for training deep neural networks that contain many layers of non-linear hidden units and a very large output layer.

Deep learning often uses convolutional neural networks (CNNs), whose origins can be traced back to the Neocognitron introduced by Kunihiko Fukushima in 1980. In 1989, Yann LeCun and colleagues applied backpropagation to such an architecture. In the early 2000s, in an industrial application CNNs already processed an estimated 10% to 20% of all the checks written in the US. Since 2011, fast implementations of CNNs on GPUs have won many visual pattern recognition competitions.

Deep feedforward neural networks were used in conjunction with reinforcement learning by AlphaGo, Google Deepmind's program that was the first to beat a professional human player.

Deep Recurrent Neural Networks

Early on, deep learning was also applied to sequence learning with recurrent neural networks (RNNs) which are general computers and can run arbitrary programs to process arbitrary sequences of inputs. The depth of an RNN is unlimited and depends on the length of its input sequence. RNNs can be trained by gradient descent but suffer from the vanishing gradient problem. In 1992, it was shown that unsupervised pre-training of a stack of recurrent neural networks can speed up subsequent supervised learning of deep sequential problems.

Numerous researchers now use variants of a deep learning recurrent NN called the Long short term memory (LSTM) network published by Hochreiter & Schmidhuber in 1997. LSTM is often trained by Connectionist Temporal Classification (CTC). At Google, Microsoft and Baidu this approach has revolutionised speech recognition. For example, in 2015, Google's speech recognition experienced a dramatic performance jump of 49% through CTC-trained LSTM, which is now available through Google Voice to billions of smartphone users. Google also used LSTM to improve machine translation, Language Modeling and Multilingual Language Processing. LSTM combined with CNNs also improved automatic image captioning and a plethora of other applications.

Control Theory

Control theory, the grandchild of cybernetics, has many important applications, especially in robotics.

Languages

AI researchers have developed several specialized languages for AI research, including Lisp and Prolog.

Evaluating Progress

In 1950, Alan Turing proposed a general procedure to test the intelligence of an agent now known as the Turing test. This procedure allows almost all the major problems of artificial intelligence to be tested. However, it is a very difficult challenge and at present all agents fail.

Artificial intelligence can also be evaluated on specific problems such as small problems in chemistry, hand-writing recognition and game-playing. Such tests have been termed subject matter expert Turing tests. Smaller problems provide more achievable goals and there are an ever-increasing number of positive results.

One classification for outcomes of an AI test is:

1. Optimal: it is not possible to perform better.

2. Strong super-human: performs better than all humans.

3. Super-human: performs better than most humans.

4. Sub-human: performs worse than most humans.

For example, performance at draughts (i.e. checkers) is optimal, performance at chess is super-human and nearing strong super-human and performance at many everyday tasks (such as recognizing a face or crossing a room without bumping into something) is sub-human.

A quite different approach measures machine intelligence through tests which are developed from *mathematical* definitions of intelligence. Examples of these kinds of tests start in the late nineties devising intelligence tests using notions from Kolmogorov complexity and data compression. Two major advantages of mathematical definitions are their applicability to nonhuman intelligences and their absence of a requirement for human testers.

A derivative of the Turing test is the Completely Automated Public Turing test to tell Computers and Humans Apart (CAPTCHA). As the name implies, this helps to determine that a user is an actual person and not a computer posing as a human. In contrast to the standard Turing test, CAPTCHA administered by a machine and targeted to a human as opposed to being administered by a human and targeted to a machine. A computer asks a user to complete a simple test then generates a grade for that test. Computers are unable to solve the problem, so correct solutions are deemed to be the result of a person taking the test. A common type of CAPTCHA is the test that requires the typing of distorted letters, numbers or symbols that appear in an image undecipherable by a computer.

Applications

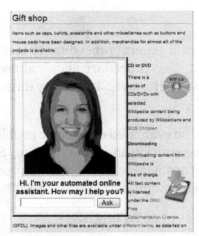

An automated online assistant providing customer service on a web page – one of many very primitive applications of artificial intelligence.

AI is relevant to any intellectual task. Modern artificial intelligence techniques are pervasive and are too numerous to list here. Frequently, when a technique reaches mainstream use, it is no longer considered artificial intelligence; this phenomenon is described as the AI effect.

High-profile examples of AI include autonomous vehicles (such as drones and self-driving cars), medical diagnosis, creating art (such as poetry), proving mathemetical theorems, playing games (such as Chess or Go), search engines (such as Google search), online assistants (such as Siri), image recognition in photographs, spam filtering, and targeting online advertisements.

With social media sites overtaking TV as a source for news for young people and news organisations increasingly reliant on social media platforms for generating distribution, major publishers now use artificial intelligence (AI) technology to post stories more effectively and generate higher volumes of traffic.

Competitions and Prizes

There are a number of competitions and prizes to promote research in artificial intelligence. The main areas promoted are: general machine intelligence, conversational behavior, data-mining, robotic cars, robot soccer and games.

Platforms

A platform (or "computing platform") is defined as "some sort of hardware architecture or software framework (including application frameworks), that allows software to run." As Rodney Brooks pointed out many years ago, it is not just the artificial intelligence software that defines the AI features of the platform, but rather the actual platform itself that affects the AI that results, i.e., there needs to be work in AI problems on real-world platforms rather than in isolation.

A wide variety of platforms has allowed different aspects of AI to develop, ranging from expert systems such as Cyc to deep-learning frameworks to robot platforms such as the Roomba with open interface. Recent advances in deep artificial neural networks and distributed computing have led to a proliferation of software libraries, including Deeplearning4j, TensorFlow, Theano and Torch.

Philosophy and Ethics

There are three philosophical questions related to AI:

1. Is artificial general intelligence possible? Can a machine solve any problem that a human being can solve using intelligence? Or are there hard limits to what a machine can accomplish?

2. Are intelligent machines dangerous? How can we ensure that machines behave ethically and that they are used ethically?

3. Can a machine have a mind, consciousness and mental states in exactly the same sense that human beings do? Can a machine be sentient, and thus deserve certain rights? Can a machine intentionally cause harm?

The Limits of Artificial General Intelligence

Can a machine be intelligent? Can it "think"?

Turing's "polite convention"

> We need not decide if a machine can "think"; we need only decide if a machine can act as intelligently as a human being. This approach to the philosophical problems associated with artificial intelligence forms the basis of the Turing test.

The Dartmouth proposal

> "Every aspect of learning or any other feature of intelligence can be so precisely described that a machine can be made to simulate it." This conjecture was printed in the proposal for the Dartmouth Conference of 1956, and represents the position of most working AI researchers.

Newell and Simon's physical symbol system hypothesis

> "A physical symbol system has the necessary and sufficient means of general

intelligent action." Newell and Simon argue that intelligence consists of formal operations on symbols. Hubert Dreyfus argued that, on the contrary, human expertise depends on unconscious instinct rather than conscious symbol manipulation and on having a "feel" for the situation rather than explicit symbolic knowledge.

Gödelian arguments

Gödel himself, John Lucas (in 1961) and Roger Penrose (in a more detailed argument from 1989 onwards) made highly technical arguments that human mathematicians can consistently see the truth of their own "Gödel statements" and therefore have computational abilities beyond that of mechanical Turing machines. However, the modern consensus in the scientific and mathematical community is that these "Gödelian arguments" fail.

The artificial brain argument

The brain can be simulated by machines and because brains are intelligent, simulated brains must also be intelligent; thus machines can be intelligent. Hans Moravec, Ray Kurzweil and others have argued that it is technologically feasible to copy the brain directly into hardware and software, and that such a simulation will be essentially identical to the original.

The AI effect

Machines are *already* intelligent, but observers have failed to recognize it. When Deep Blue beat Garry Kasparov in chess, the machine was acting intelligently. However, onlookers commonly discount the behavior of an artificial intelligence program by arguing that it is not "real" intelligence after all; thus "real" intelligence is whatever intelligent behavior people can do that machines still can not. This is known as the AI Effect: "AI is whatever hasn't been done yet."

Intelligent Behaviour and Machine Ethics

As a minimum, an AI system must be able to reproduce aspects of human intelligence. This raises the issue of how ethically the machine should behave towards both humans and other AI agents. This issue was addressed by Wendell Wallach in his book titled *Moral Machines* in which he introduced the concept of artificial moral agents (AMA). For Wallach, AMAs have become a part of the research landscape of artificial intelligence as guided by its two central questions which he identifies as "Does Humanity Want Computers Making Moral Decisions" and "Can (Ro)bots Really Be Moral". For Wallach the question is not centered on the issue of *whether* machines can demonstrate the equivalent of moral behavior in contrast to the *constraints* which society may place on the development of AMAs.

Machine Ethics

The field of machine ethics is concerned with giving machines ethical principles, or a procedure for discovering a way to resolve the ethical dilemmas they might encounter, enabling them to function in an ethically responsible manner through their own ethical decision making. The field was delineated in the AAAI Fall 2005 Symposium on Machine Ethics: "Past research concerning the relationship between technology and ethics has largely focused on responsible and irresponsible use of technology by human beings, with a few people being interested in how human beings ought to treat machines. In all cases, only human beings have engaged in ethical reasoning. The time has come for adding an ethical dimension to at least some machines. Recognition of the ethical ramifications of behavior involving machines, as well as recent and potential developments in machine autonomy, necessitate this. In contrast to computer hacking, software property issues, privacy issues and other topics normally ascribed to computer ethics, machine ethics is concerned with the behavior of machines towards human users and other machines. Research in machine ethics is key to alleviating concerns with autonomous systems—it could be argued that the notion of autonomous machines without such a dimension is at the root of all fear concerning machine intelligence. Further, investigation of machine ethics could enable the discovery of problems with current ethical theories, advancing our thinking about Ethics." Machine ethics is sometimes referred to as machine morality, computational ethics or computational morality. A variety of perspectives of this nascent field can be found in the collected edition "Machine Ethics" that stems from the AAAI Fall 2005 Symposium on Machine Ethics. Some suggest that to ensure that AI-equipped machines (sometimes called "smart machines") will act ethically requires a new kind of AI. This AI would be able to monitor, supervise, and if need be, correct the first order AI.

Malevolent and Friendly AI

Political scientist Charles T. Rubin believes that AI can be neither designed nor guaranteed to be benevolent. He argues that "any sufficiently advanced benevolence may be indistinguishable from malevolence." Humans should not assume machines or robots would treat us favorably, because there is no *a priori* reason to believe that they would be sympathetic to our system of morality, which has evolved along with our particular biology (which AIs would not share). Hyper-intelligent software may not necessarily decide to support the continued existence of mankind, and would be extremely difficult to stop. This topic has also recently begun to be discussed in academic publications as a real source of risks to civilization, humans, and planet Earth.

Physicist Stephen Hawking, Microsoft founder Bill Gates and SpaceX founder Elon Musk have expressed concerns about the possibility that AI could evolve to the point that humans could not control it, with Hawking theorizing that this could "spell the end of the human race".

One proposal to deal with this is to ensure that the first generally intelligent AI is 'Friendly AI', and will then be able to control subsequently developed AIs. Some question whether this kind of check could really remain in place.

Leading AI researcher Rodney Brooks writes, "I think it is a mistake to be worrying about us developing malevolent AI anytime in the next few hundred years. I think the worry stems from a fundamental error in not distinguishing the difference between the very real recent advances in a particular aspect of AI, and the enormity and complexity of building sentient volitional intelligence."

Devaluation of Humanity

Joseph Weizenbaum wrote that AI applications can not, by definition, successfully simulate genuine human empathy and that the use of AI technology in fields such as customer service or psychotherapy was deeply misguided. Weizenbaum was also bothered that AI researchers (and some philosophers) were willing to view the human mind as nothing more than a computer program (a position now known as computationalism). To Weizenbaum these points suggest that AI research devalues human life.

Decrease in Demand for Human Labor

Martin Ford, author of *The Lights in the Tunnel: Automation, Accelerating Technology and the Economy of the Future,* and others argue that specialized artificial intelligence applications, robotics and other forms of automation will ultimately result in significant unemployment as machines begin to match and exceed the capability of workers to perform most routine and repetitive jobs. Ford predicts that many knowledge-based occupations—and in particular entry level jobs—will be increasingly susceptible to automation via expert systems, machine learning and other AI-enhanced applications. AI-based applications may also be used to amplify the capabilities of low-wage offshore workers, making it more feasible to outsource knowledge work.

Machine Consciousness, Sentience and Mind

If an AI system replicates all key aspects of human intelligence, will that system also be sentient – will it have a mind which has conscious experiences? This question is closely related to the philosophical problem as to the nature of human consciousness, generally referred to as the hard problem of consciousness.

Consciousness

Computationalism

Computationalism is the position in the philosophy of mind that the human mind or the human brain (or both) is an information processing system and that thinking is a form of computing. Computationalism argues that the relationship between mind and body is similar or identical to the relationship between software and hardware and thus may be a solution to the mind-body problem. This philosophical position was inspired by the work of AI researchers and cognitive scientists in the 1960s and was originally proposed by philosophers Jerry Fodor and Hilary Putnam.

Strong AI Hypothesis

The philosophical position that John Searle has named "strong AI" states: "The appropriately programmed computer with the right inputs and outputs would thereby have a mind in exactly the same sense human beings have minds." Searle counters this assertion with his Chinese room argument, which asks us to look *inside* the computer and try to find where the "mind" might be.

Robot Rights

Mary Shelley's *Frankenstein* considers a key issue in the ethics of artificial intelligence: if a machine can be created that has intelligence, could it also *feel*? If it can feel, does it have the same rights as a human? The idea also appears in modern science fiction, such as the film *A.I.: Artificial Intelligence*, in which humanoid machines have the ability to feel emotions. This issue, now known as "robot rights", is currently being considered by, for example, California's Institute for the Future, although many critics believe that the discussion is premature. The subject is profoundly discussed in the 2010 documentary film *Plug & Pray*.

Superintelligence

Are there limits to how intelligent machines – or human-machine hybrids – can be? A superintelligence, hyperintelligence, or superhuman intelligence is a hypothetical agent that would possess intelligence far surpassing that of the brightest and most gifted human mind. "Superintelligence" may also refer to the form or degree of intelligence possessed by such an agent.

Technological Singularity

If research into Strong AI produced sufficiently intelligent software, it might be able to reprogram and improve itself. The improved software would be even better at improving itself, leading to recursive self-improvement. The new intelligence could thus increase exponentially and dramatically surpass humans. Science fiction writer Vernor Vinge named this scenario "singularity". Technological singularity is when accelerating progress in technologies will cause a runaway effect wherein artificial intelligence will exceed human intellectual capacity and control, thus radically changing or even ending civilization. Because the capabilities of such an intelligence may be impossible to comprehend, the technological singularity is an occurrence beyond which events are unpredictable or even unfathomable.

Ray Kurzweil has used Moore's law (which describes the relentless exponential improvement in digital technology) to calculate that desktop computers will have the same processing power as human brains by the year 2029, and predicts that the singularity will occur in 2045.

Transhumanism

You awake one morning to find your brain has another lobe functioning. Invisible, this auxiliary lobe answers your questions with information beyond the realm of your own memory, suggests plausible courses of action, and asks questions that help bring out relevant facts. You quickly come to rely on the new lobe so much that you stop wondering how it works. You just use it. This is the dream of artificial intelligence.

—BYTE, April 1985

Robot designer Hans Moravec, cyberneticist Kevin Warwick and inventor Ray Kurzweil have predicted that humans and machines will merge in the future into cyborgs that are more capable and powerful than either. This idea, called transhumanism, which has roots in Aldous Huxley and Robert Ettinger, has been illustrated in fiction as well, for example in the manga *Ghost in the Shell* and the science-fiction series *Dune*.

In the 1980s artist Hajime Sorayama's Sexy Robots series were painted and published in Japan depicting the actual organic human form with lifelike muscular metallic skins and later "the Gynoids" book followed that was used by or influenced movie makers including George Lucas and other creatives. Sorayama never considered these organic robots to be real part of nature but always unnatural product of the human mind, a fantasy existing in the mind even when realized in actual form.

Edward Fredkin argues that "artificial intelligence is the next stage in evolution", an idea first proposed by Samuel Butler's "Darwin among the Machines" (1863), and expanded upon by George Dyson in his book of the same name in 1998.

Existential Risk

The development of full artificial intelligence could spell the end of the human race. Once humans develop artificial intelligence, it will take off on its own and redesign itself at an ever-increasing rate. Humans, who are limited by slow biological evolution, couldn't compete and would be superseded.

—Stephen Hawking

A common concern about the development of artificial intelligence is the potential threat it could pose to mankind. This concern has recently gained attention after mentions by celebrities including Stephen Hawking, Bill Gates, and Elon Musk. A group of prominent tech titans including Peter Thiel, Amazon Web Services and Musk have committed $1billion to OpenAI a nonprofit company aimed at championing responsible AI development. The opinion of experts within the field of artificial intelligence is mixed, with sizable fractions both concerned and unconcerned by risk from eventual superhumanly-capable AI.

In his book *Superintelligence*, Nick Bostrom provides an argument that artificial intelligence will pose a threat to mankind. He argues that sufficiently intelligent AI, if it chooses actions based on achieving some goal, will exhibit convergent behavior such as acquiring resources or protecting itself from being shut down. If this AI's goals do not reflect humanity's - one example is an AI told to compute as many digits of pi as possible - it might harm humanity in order to acquire more resources or prevent itself from being shut down, ultimately to better achieve its goal.

For this danger to be realized, the hypothetical AI would have to overpower or out-think all of humanity, which a minority of experts argue is a possibility far enough in the future to not be worth researching. Other counterarguments revolve around humans being either intrinsically or convergently valuable from the perspective of an artificial intelligence.

Concern over risk from artificial intelligence has led to some high-profile donations and investments. In January 2015, Elon Musk donated ten million dollars to the Future of Life Institute to fund research on understanding AI decision making. The goal of the institute is to "grow wisdom with which we manage" the growing power of technology. Musk also funds companies developing artificial intelligence such as Google DeepMind and Vicarious to "just keep an eye on what's going on with artificial intelligence. I think there is potentially a dangerous outcome there."

Development of militarized artificial intelligence is a related concern. Currently, 50+ countries are researching battlefield robots, including the United States, China, Russia, and the United Kingdom. Many people concerned about risk from superintelligent AI also want to limit the use of artificial soldiers.

Moral Decision-making

To keep AI ethical, some have suggested teaching new technologies equipped with AI, such as driver-less cars, to render moral decisions on their own. Others argued that these technologies could learn to act ethically the way children do—by interacting with adults, in particular, with ethicists. Still others suggest these smart technologies can determine the moral preferences of those who use them (just the way one learns about consumer preferences) and then be programmed to heed these preferences.

In Fiction

The implications of artificial intelligence have been a persistent theme in science fiction. Early stories typically revolved around intelligent robots. The word "robot" itself was coined by Karel Čapek in his 1921 play *R.U.R.*, the title standing for "Rossum's Universal Robots". Later, the SF writer Isaac Asimov developed the Three Laws of Robotics which he subsequently explored in a long series of robot stories. These laws have since gained some traction in genuine AI research.

Other influential fictional intelligences include HAL, the computer in charge of the spaceship in *2001: A Space Odyssey*, released as both a film and a book in 1968 and written by Arthur C. Clarke.

AI has since become firmly rooted in popular culture and is in many films, such as *The Terminator* (1984) and *A.I. Artificial Intelligence* (2001).

Artificial Consciousness

Artificial consciousness (AC), also known as machine consciousness (MC) or synthetic consciousness (Gamez 2008; Reggia 2013), is a field related to artificial intelligence and cognitive robotics. The aim of the theory of artificial consciousness is to "define that which would have to be synthesized were consciousness to be found in an engineered artifact" (Aleksander 1995).

Neuroscience hypothesizes that consciousness is generated by the interoperation of various parts of the brain, called the neural correlates of consciousness or NCC, though there are challenges to that perspective. Proponents of AC believe it is possible to construct systems (e.g., computer systems) that can emulate this NCC interoperation.

Artificial consciousness concepts are also pondered in the philosophy of artificial intelligence through questions about mind, consciousness, and mental states.

Philosophical Views

As there are many hypothesized types of consciousness, there are many potential implementations of artificial consciousness. In the philosophical literature, perhaps the most common taxonomy of consciousness is into "access" and "phenomenal" variants. Access consciousness concerns those aspects of experience that can be apprehended, while phenomenal consciousness concerns those aspects of experience that seemingly cannot be apprehended, instead being characterized qualitatively in terms of "raw feels", "what it is like" or qualia (Block 1997).

Plausibility Debate

Type-identity theorists and other skeptics hold the view that consciousness can only be realized in particular physical systems because consciousness has properties that necessarily depend on physical constitution (Block 1978; Bickle 2003).

In his article "Artificial Consciousness: Utopia or Real Possibility" Giorgio Buttazzo says that despite our current technology's ability to simulate autonomy, "Working in a fully automated mode, they [the computers] cannot exhibit creativity, emotions, or free will. A computer, like a washing machine, is a slave operated by its components."

For other theorists (e.g., functionalists), who define mental states in terms of causal roles, any system that can instantiate the same pattern of causal roles, regardless of physical constitution, will instantiate the same mental states, including consciousness (Putnam 1967).

Computational Foundation Argument

One of the most explicit arguments for the plausibility of AC comes from David Chalmers. His proposal, found within his article Chalmers 2011, is roughly that the right kinds of computations are sufficient for the possession of a conscious mind. In the outline, he defends his claim thus: Computers perform computations. Computations can capture other systems' abstract causal organization.

The most controversial part of Chalmers' proposal is that mental properties are "organizationally invariant". Mental properties are of two kinds, psychological and phenomenological. Psychological properties, such as belief and perception, are those that are "characterized by their causal role". He adverts to the work of Armstrong 1968 and Lewis 1972 in claiming that "[s]ystems with the same causal topology...will share their psychological properties."

Phenomenological properties are not prima facie definable in terms of their causal roles. Establishing that phenomenological properties are amenable to individuation by causal role therefore requires argument. Chalmers provides his Dancing Qualia Argument for this purpose.

Chalmers begins by assuming that agents with identical causal organizations could have different experiences. He then asks us to conceive of changing one agent into the other by the replacement of parts (neural parts replaced by silicon, say) while preserving its causal organization. Ex hypothesi, the experience of the agent under transformation would change (as the parts were replaced), but there would be no change in causal topology and therefore no means whereby the agent could "notice" the shift in experience.

Critics of AC object that Chalmers begs the question in assuming that all mental properties and external connections are sufficiently captured by abstract causal organization.

Ethics

If it were certain that a particular machine was conscious, its rights would be an ethical issue that would need to be assessed (e.g. what rights it would have under law). For example, a conscious computer that was owned and used as a tool or central computer of a building or large machine is a particular ambiguity. Should laws be made for such a case, consciousness would also require a legal definition (for example a machine's ability to experience pleasure or pain, known as sentience). Because artificial consciousness is still largely a theoretical subject, such ethics have not been discussed or developed to a great extent, though it has often been a theme in fiction.

The rules for the 2003 Loebner Prize competition explicitly addressed the question of robot rights:

61. If, in any given year, a publicly available open source Entry entered by the University of Surrey or the Cambridge Center wins the Silver Medal or the Gold Medal, then the Medal and the Cash Award will be awarded to the body responsible for the development of that Entry. If no such body can be identified, or if there is disagreement among two or more claimants, the Medal and the Cash Award will be held in trust *until such time as the Entry may legally possess, either in the United States of America or in the venue of the contest, the Cash Award and Gold Medal in its own right.*

Research and Implementation Proposals

Aspects of Consciousness

There are various aspects of consciousness generally deemed necessary for a machine to be artificially conscious. A variety of functions in which consciousness plays a role were suggested by Bernard Baars (Baars 1988) and others. The functions of consciousness suggested by Bernard Baars are Definition and Context Setting, Adaptation and Learning, Editing, Flagging and Debugging, Recruiting and Control, Prioritizing and Access-Control, Decision-making or Executive Function, Analogy-forming Function, Metacognitive and Self-monitoring Function, and Autoprogramming and Self-maintenance Function. Igor Aleksander suggested 12 principles for artificial consciousness (Aleksander 1995) and these are: The Brain is a State Machine, Inner Neuron Partitioning, Conscious and Unconscious States, Perceptual Learning and Memory, Prediction, The Awareness of Self, Representation of Meaning, Learning Utterances, Learning Language, Will, Instinct, and Emotion. The aim of AC is to define whether and how these and other aspects of consciousness can be synthesized in an engineered artifact such as a digital computer. This list is not exhaustive; there are many others not covered.

Awareness

Awareness could be one required aspect, but there are many problems with the exact definition of *awareness*. The results of the experiments of neuroscanning on monkeys suggest that a process, not only a state or object, activates neurons. Awareness includes creating and testing alternative models of each process based on the information received through the senses or imagined, and is also useful for making predictions. Such modeling needs a lot of flexibility. Creating such a model includes modeling of the physical world, modeling of one's own internal states and processes, and modeling of other conscious entities.

There are at least three types of awareness: agency awareness, goal awareness, and sensorimotor awareness, which may also be conscious or not. For example, in agency awareness you may be aware that you performed a certain action yesterday, but are not now conscious of it. In goal awareness you may be aware that you must search for a lost object, but are not now conscious of it. In sensorimotor awareness, you may be aware that your hand is resting on an object, but are not now conscious of it.

Because objects of awareness are often conscious, the distinction between awareness and consciousness is frequently blurred or they are used as synonyms.

Memory

Conscious events interact with memory systems in learning, rehearsal, and retrieval. The IDA model elucidates the role of consciousness in the updating of perceptual memory, transient episodic memory, and procedural memory. Transient episodic and declarative memories have distributed representations in IDA, there is evidence that this is also the case in the nervous system. In IDA, these two memories are implemented computationally using a modified version of Kanerva's Sparse distributed memory architecture.

Learning

Learning is also considered necessary for AC. By Bernard Baars, conscious experience is needed to represent and adapt to novel and significant events (Baars 1988). By Axel Cleeremans and Luis Jiménez, learning is defined as "a set of philogenetically [sic] advanced adaptation processes that critically depend on an evolved sensitivity to subjective experience so as to enable agents to afford flexible control over their actions in complex, unpredictable environments" (Cleeremans 2001).

Anticipation

The ability to predict (or anticipate) foreseeable events is considered important for AC by Igor Aleksander. The emergentist multiple drafts principle proposed by Daniel Dennett in *Consciousness Explained* may be useful for prediction: it involves the evaluation and selection of the most appropriate "draft" to fit the current environment. Anticipation includes prediction of consequences of one's own proposed actions and prediction of consequences of probable actions by other entities.

Relationships between real world states are mirrored in the state structure of a conscious organism enabling the organism to predict events. An artificially conscious machine should be able to anticipate events correctly in order to be ready to respond to them when they occur or to take premptive action to avert anticipated events. The implication here is that the machine needs flexible, real-time components that build spatial, dynamic, statistical, functional, and cause-effect models of the real world and predicted worlds, making it possible to demonstrate that it possesses artificial consciousness in the present and future and not only in the past. In order to do this, a conscious machine should make coherent predictions and contingency plans, not only in worlds with fixed rules like a chess board, but also for novel environments that may change, to be executed only when appropriate to simulate and control the real world.

Subjective Experience

Subjective experiences or qualia are widely considered to be *the* hard problem of consciousness. Indeed, it is held to pose a challenge to physicalism, let alone computationalism. On the other hand, there are problems in other fields of science which limit that which we can observe, such as the uncertainty principle in physics, which have not made the research in these fields of science impossible.

Role of Cognitive Architectures

The term "cognitive architecture" may refer to a theory about the structure of the human mind, or any portion or function thereof, including consciousness. In another context, a cognitive architecture implements the theory on computers. An example is *QuBIC: Quantum and Bio-inspired Cognitive Architecture for Machine Consciousness*. One of the main goals of a cognitive architecture is to summarize the various results of cognitive psychology in a comprehensive computer model. However, the results need to be in a formalized form so they can be the basis of a computer program.

Symbolic or Hybrid Proposals

Franklin's Intelligent Distribution Agent

Stan Franklin (1995, 2003) defines an autonomous agent as possessing functional consciousness when it is capable of several of the functions of consciousness as identified by Bernard Baars' Global Workspace Theory (Baars 1988, 1997). His brain child IDA (Intelligent Distribution Agent) is a software implementation of GWT, which makes it functionally conscious by definition. IDA's task is to negotiate new assignments for sailors in the US Navy after they end a tour of duty, by matching each individual's skills and preferences with the Navy's needs. IDA interacts with Navy databases and communicates with the sailors via natural language e-mail dialog while obeying a large set of Navy policies. The IDA computational model was developed during 1996–2001 at Stan Franklin's "Conscious" Software Research Group at the University of Memphis. It "consists of approximately a quarter-million lines of Java code, and almost completely consumes the resources of a 2001 high-end workstation." It relies heavily on *codelets*, which are "special purpose, relatively independent, mini-agent[s] typically implemented as a small piece of code running as a separate thread." In IDA's top-down architecture, high-level cognitive functions are explicitly modeled. While IDA is functionally conscious by definition, Franklin does "not attribute phenomenal consciousness to his own 'conscious' software agent, IDA, in spite of her many human-like behaviours. This in spite of watching several US Navy detailers repeatedly nodding their heads saying 'Yes, that's how I do it' while watching IDA's internal and external actions as she performs her task."

Ron Sun's Cognitive Architecture CLARION

CLARION posits a two-level representation that explains the distinction between conscious and unconscious mental processes.

CLARION has been successful in accounting for a variety of psychological data. A number of well-known skill learning tasks have been simulated using CLARION that span the spectrum ranging from simple reactive skills to complex cognitive skills. The tasks include serial reaction time (SRT) tasks, artificial grammar learning (AGL) tasks, process control (PC) tasks, the categorical inference (CI) task, the alphabetical arithmetic (AA) task, and the Tower of Hanoi (TOH) task (Sun 2002). Among them, SRT, AGL, and PC are typical implicit learning tasks, very much relevant to the issue of consciousness as they operationalized the notion of consciousness in the context of psychological experiments.

Ben Goertzel's OpenCog

Ben Goertzel is pursuing an embodied AGI through the open-source OpenCog project. Current code includes embodied virtual pets capable of learning simple English-language commands, as well as integration with real-world robotics, being done at the Hong Kong Polytechnic University.

Connectionist Proposals

Haikonen's Cognitive Architecture

Pentti Haikonen (2003) considers classical rule-based computing inadequate for achieving AC: "the brain is definitely not a computer. Thinking is not an execution of programmed strings of commands. The brain is not a numerical calculator either. We do not think by numbers." Rather than trying to achieve mind and consciousness by identifying and implementing their underlying computational rules, Haikonen proposes "a special cognitive architecture to reproduce the processes of perception, inner imagery, inner speech, pain, pleasure, emotions and the cognitive functions behind these. This bottom-up architecture would produce higher-level functions by the power of the elementary processing units, the artificial neurons, without algorithms or programs". Haikonen believes that, when implemented with sufficient complexity, this architecture will develop consciousness, which he considers to be "a style and way of operation, characterized by distributed signal representation, perception process, cross-modality reporting and availability for retrospection." Haikonen is not alone in this process view of consciousness, or the view that AC will spontaneously emerge in autonomous agents that have a suitable neuro-inspired architecture of complexity; these are shared by many, e.g. Freeman (1999) and Cotterill (2003). A low-complexity implementation of the architecture proposed by Haikonen (2003) was reportedly not capable of AC, but did exhibit emotions as expected. See Doan (2009) for a comprehensive introduction to Haikonen's cognitive architecture. An updated account of Haikonen's architecture, along with a summary of his philosophical views, is given in Haikonen (2012).

Shanahan's Cognitive Architecture

Murray Shanahan describes a cognitive architecture that combines Baars's idea of a global workspace with a mechanism for internal simulation ("imagination") (Shanah-an 2006).

Takeno's Self-awareness Research

Self-awareness in robots is being investigated by Junichi Takeno at Meiji University in Japan. Takeno is asserting that he has developed a robot capable of discriminating between a self-image in a mirror and any other having an identical image to it, and this claim has already been reviewed (Takeno, Inaba & Suzuki 2005). Takeno asserts that he first contrived the computational module called a MoNAD, which has a self-aware function, and he then constructed the artificial consciousness system by formulating the relationships between emotions, feelings and reason by connecting the modules in a hierarchy (Igarashi, Takeno 2007). Takeno completed a mirror image cognition experiment using a robot equipped with the MoNAD system. Takeno proposed the Self-Body Theory stating that "humans feel that their own mirror image is closer to themselves than an actual part of themselves." The most important point in developing artificial consciousness or clarifying human consciousness is the development of a function of self awareness, and he claims that he has demonstrated physical and mathematical evidence for this in his thesis. He also demonstrated that robots can study episodes in memory where the emotions were stimulated and use this experience to take predictive actions to prevent the recurrence of unpleasant emotions (Torigoe, Takeno 2009).

Aleksander's Impossible Mind

Igor Aleksander, emeritus professor of Neural Systems Engineering at Imperial College, has extensively researched artificial neural networks and claims in his book *Impossible Minds: My neurons, My Consciousness* that the principles for creating a conscious machine already exist but that it would take forty years to train such a machine to understand language. Whether this is true remains to be demonstrated and the basic principle stated in *Impossible minds*—that the brain is a neural state machine—is open to doubt.

Thaler's Creativity Machine Paradigm

Stephen Thaler proposed a possible connection between consciousness and creativity in his 1994 patent, called "Device for the Autonomous Generation of Useful Information" (DAGUI), or the so-called "Creativity Machine," in which computational critics govern the injection of synaptic noise and degradation into neural nets so as to induce false memories or confabulations that may qualify as potential ideas or strategies. He recruits this neural architecture and methodology to account for the subjective feel of consciousness, claiming that similar noise-driven neural assemblies within the brain invent dubious significance to overall cortical activity. Thaler's theory and the resulting patents in machine consciousness were inspired by experiments in which he internally disrupted trained neural nets so as to drive a succession of neural activation patterns that he likened to stream of consciousness.

Michael Graziano's Attention Schema

In 2011, Michael Graziano and Sabine Kastler published a paper named "Human consciousness and its relationship to social neuroscience: A novel hypothesis" proposing a theory of consciousness as an attention schema. Graziano went on to publish an expanded discussion of this theory in his book "Consciousness and the Social Brain". This Attention Schema Theory of Consciousness, as he named it, proposes that the brain tracks attention to various sensory inputs by way of an attention schema, analogous to the well study body schema that tracks the spatial place of a person's body. This relates to artificial consciousness by proposing a specific mechanism of information handling, that produces what we allegedly experience and describe as consciousness, and which should be able to be duplicated by a machine using current technology. When the brain finds that person X is aware of thing Y, it is in effect modeling the state in which person X is applying an attentional enhancement to Y. In the attention schema theory, the same process can be applied to oneself. The brain tracks attention to various sensory inputs, and one's own awareness is a schematized model of one's attention. Graziano proposes specific locations in the brain for this process, and suggests that such awareness is a computed feature constructed by an expert system in the brain.

Testing

The most well-known method for testing machine intelligence is the Turing test. But when interpreted as only observational, this test contradicts the philosophy of science principles of theory dependence of observations. It also has been suggested that Alan Turing's recommendation of imitating not a human adult consciousness, but a human child consciousness, should be taken seriously.

Other tests, such as ConsScale, test the presence of features inspired by biological systems, or measure the cognitive development of artificial systems.

Qualia, or phenomenological consciousness, is an inherently first-person phenomenon. Although various systems may display various signs of behavior correlated with functional consciousness, there is no conceivable way in which third-person tests can have access to first-person phenomenological features. Because of that, and because there is no empirical definition of consciousness, a test of presence of consciousness in AC may be impossible.

In Fiction

Characters with artificial consciousness (or at least with personalities that imply they have consciousness), from works of fiction:

- AC created by merging 2 AIs in the *Sprawl trilogy* by William Gibson

- Agents in the simulated reality known as "The Matrix" in *The Matrix* franchise

  o Agent Smith, began as an Agent in *The Matrix*, then became a renegade program of overgrowing power that could make copies of itself like a self-replicating computer virus

- AM (Allied Mastercomputer), the antagonist of *Harlan Ellison*'s short novel *I Have No Mouth, and I Must Scream*

- Amusement park robots (with pixilated consciousness) that went homicidal in *Westworld* and *Futureworld*

- Arnold Rimmer computer-generated sapient hologram, aboard the *Red Dwarf* deep space ore hauler

- Ava, a humanoid robot in Ex Machina

- Ash, android crew member of the Nostromo starship in the movie *Alien*

- *The Bicentennial Man*, an android in Isaac Asimov's *Foundation* universe

- Bishop, android crew member aboard the U.S.S. Sulaco in the movie *Aliens*

- The uploaded mind of Dr. Will Caster, which presumably included his consciousness, from the film *Transcendence*

- C-3PO, protocol droid featured in all the *Star Wars* movies

- Chappie in the movie *CHAPPiE*

- Cohen and other Emergent AIs in Chris Moriarty's *Spin* Series

- Commander Data in *Star Trek: The Next Generation*

- Cortana and other "Smart AI" from the *Halo* series of games

- Cylons, genocidal robots with resurrection ships that enable the consciousness of any Cylon within an unspecified range to download into a new body aboard the ship upon death. From *Battlestar Galactica*.

- Erasmus, baby killer robot that incited the Butlerian Jihad in the *Dune* franchise

- The Geth in *Mass Effect*

- HAL 9000, the paranoid computer in *2001: A Space Odyssey*

- Holly, ship's computer with an IQ of 6000, aboard the *Red Dwarf*

- Jane in Orson Scott Card's *Speaker for the Dead, Xenocide, Children of the Mind*, and *Investment Counselor*

- Johnny Five from the movie *Short Circuit*

- Joshua from the movie *War Games*

- Keymaker, an "exile" sapient program in *The Matrix* franchise

- "Machine" – android from the film *The Machine*, whose owners try to kill her when they witness her conscious thoughts, out of fear that she will design better androids (intelligence explosion)

- Mike, from The Moon Is a Harsh Mistress

- Mimi, humanoid robot in Real Humans - "Äkta människor" (original title) 2012

- The Minds in Iain M. Banks' *Culture* novels.

- Omnius, sentient computer network that controlled the Universe until overthrown by the Butlerian Jihad in the *Dune* franchise

- Operating Systems in the movie *Her*

- The Oracle, sapient program in *The Matrix* franchise

- The sentient holodeck character Professor James Moriarty in the *Ship in a Bottle* episode from *Star Trek: The Next Generation*

- In Greg Egan's novel *Permutation City* the protagonist creates digital copies of himself to conduct experiments that are also related to implications of artificial consciousness on identity

- Puppet Master in *Ghost in the Shell* manga and anime

- R2-D2, exciteable astromech droid featured in all the *Star Wars* movies

- Replicants, biorobotic androids from the novel *Do Androids Dream of Electric*

Sheep? and the movie *Blade Runner* which portray what might happen when artificially conscious robots are modeled very closely upon humans

- Roboduck, combat robot superhero in the *NEW-GEN* comic book series from Marvel Comics

- Robots in Isaac Asimov's *Robot* series

- Robots in *The Matrix* franchise, especially in *The Animatrix*

- The Ship (the result of a large-scale AC experiment) in Frank Herbert's *Destination: Void* and sequels, despite past edicts warning against "Making a Machine in the Image of a Man's Mind."

- Skynet from the *Terminator* franchise

- "Synths" are a type of android in the video game Fallout 4. There is a faction in the game known as "the Railroad" which believes that, as conscious beings, synths have their own rights. The Institute, the lab that produces the synths, mostly does not believe they are truly conscious and attributes any apparent desires for freedom as a malfunction.

- TARDIS, time machine and spacecraft of *Doctor Who*, sometimes portrayed with a mind of its own

- The terminator cyborgs from the *Terminator* franchise, with visual consciousness depicted via first-person perspective

- Transformers, sentient robots from the entertainment franchise of the same name

- Vanamonde in Arthur C. Clarke's *The City and the Stars*—an artificial being that was immensely powerful but entirely child-like.

- WALL-E, a robot and the title character in *WALL-E*

Neurotechnology

Neurotechnology is any technology that has a fundamental influence on how people understand the brain and various aspects of consciousness, thought, and higher order activities in the brain. It also includes technologies that are designed to improve and repair brain function and allow researchers and clinicians to visualize the brain.

Background

The field of neurotechnology has been around for nearly half a century but has only reached maturity in the last twenty years. The advent of brain imaging revolutionized the field, allowing researchers to directly monitor the brain's activities during experiments. Neurotechnology has made significant impact on society, though its presence is so commonplace that many do not realize its ubiquity. From pharmaceutical drugs to brain scanning, neurotechnology affects nearly all industrialized people either directly or indirectly, be it from drugs for depression, sleep, ADD, or anti-neurotics to cancer scanning, stroke rehabilitation, and much more.

As the field's depth increases it will potentially allow society to control and harness more of what the brain does and how it influences lifestyles and personalities. Commonplace technologies already attempt to do this; games like BrainAge, and programs like Fast ForWord that aim to improve brain function, are neurotechnologies.

Currently, modern science can image nearly all aspects of the brain as well as control a degree of the function of the brain. It can help control depression, over-activation, sleep deprivation, and many other conditions. Therapeutically it can help improve stroke victims' motor coordination, improve brain function, reduce epileptic episodes, improve patients with degenerative motor diseases (Parkinson's disease, Huntington's Disease, ALS), and can even help alleviate phantom pain perception. Advances in the field promise many new enhancements and rehabilitation methods for patients suffering from neurological problems. The neurotechnology revolution has given rise to the Decade of the Mind initiative, which was started in 2007. It also offers the possibility of revealing the mechanisms by which mind and consciousness emerge from the brain.

Current Technologies

Imaging

Magnetoencephalography is a functional neuroimaging technique for mapping brain activity by recording magnetic fields produced by electrical currents occurring naturally in the brain, using very sensitive magnetometers. Arrays of SQUIDs (superconducting quantum interference devices) are the most common magnetometer. Applications of MEG include basic research into perceptual and cognitive brain processes, localizing regions affected by pathology before surgical removal, determining the function of various parts of the brain, and neurofeedback. This can be applied in a clinical setting to find locations of abnormalities as well as in an experimental setting to simply measure brain activity.

Magnetic resonance imaging (MRI) is used for scanning the brain for topological and landmark structure in the brain, but can also be used for imaging activation in the brain. While detail about how MRI works is reserved for the actual MRI article, the uses of MRI are far reaching in the study of neuroscience. It is a cornerstone technology in studying the mind, especially with the advent of functional MRI (fMRI). Functional MRI measures the oxygen levels in the brain upon activation (higher oxygen content =

neural activation) and allows researchers to understand what loci are responsible for activation under a given stimulus. This technology is a large improvement to single cell or loci activation by means of exposing the brain and contact stimulation. Functional MRI allows researchers to draw associative relationships between different loci and regions of the brain and provides a large amount of knowledge in establishing new landmarks and loci in the brain.

Computed tomography (CT) is another technology used for scanning the brain. It has been used since the 1970s and is another tool used by neuroscientists to track brain structure and activation. While many of the functions of CT scans are now done using MRI, CT can still be used as the mode by which brain activation and brain injury are detected. Using an X-ray, researchers can detect radioactive markers in the brain that indicate brain activation as a tool to establish relationships in the brain as well as detect many injuries/diseases that can cause lasting damage to the brain such as aneurysms, degeneration, and cancer.

Positron emission tomography (PET) is another imaging technology that aids researchers. Instead of using magnetic resonance or X-rays, PET scans rely on positron emitting markers that are bound to a biologically relevant marker such as glucose. The more activation in the brain the more that region requires nutrients, so higher activation appears more brightly on an image of the brain. PET scans are becoming more frequently used by researchers because PET scans are activated due to metabolism whereas MRI is activated on a more physiological basis (sugar activation versus oxygen activation).

Transcranial Magnetic Stimulation

Transcranial magnetic stimulation (TMS) is essentially direct magnetic stimulation to the brain. Because electric currents and magnetic fields are intrinsically related, by stimulating the brain with magnetic pulses it is possible to interfere with specific loci in the brain to produce a predictable effect. This field of study is currently receiving a large amount of attention due to the potential benefits that could come out of better understanding this technology. Transcranial magnetic movement of particles in the brain shows promise for drug targeting and delivery as studies have demonstrated this to be noninvasive on brain physiology.

Transcranial Direct Current Stimulation

Transcranial direct current stimulation (tDCS) is a form of neurostimulation which uses constant, low current delivered via electrodes placed on the scalp. The mechanisms underlying tDCS effects are still incompletely understood, but recent advances in neurotechnology allowing for *in vivo* assessment of brain electric activity during tDCS promise to advance understanding of these mechanisms. Research into using tDCS on healthy adults have demonstrated that tDCS can increase cognitive performance on a variety of tasks, depending on the area of the brain being stimulated. tDCS has been used to enhance language and mathematical ability (though one form of tDCS was also found to inhibit math learning), attention span, problem solving, memory, and coordination.

Cranial Surface Measurements

Electroencephalography (EEG) is a method of measuring brainwave activity non-invasively. A number of electrodes are placed around the head and scalp and electrical signals are measured. Typically EEGs are used when dealing with sleep, as there are characteristic wave patterns associated with different stages of sleep. Clinically EEGs are used to study epilepsy as well as stroke and tumor presence in the brain. EEGs are a different method to understand the electrical signaling in the brain during activation.

Magnetoencephalography (MEG) is another method of measuring activity in the brain by measuring the magnetic fields that arise from electrical currents in the brain. The benefit to using MEG instead of EEG is that these fields are highly localized and give rise to better understanding of how specific loci react to stimulation or if these regions over-activate (as in epileptic seizures).

Implant Technologies

Neurodevices are any devices used to monitor or regulate brain activity. Currently there are a few available for clinical use as a treatment for Parkinson's disease. The most common neurodevices are deep brain stimulators (DBS) that are used to give electrical stimulation to areas stricken by inactivity. Parkinson's disease is known to be caused by an inactivation of the basal ganglia (nuclei) and recently DBS has become the more preferred form of treatment for Parkinson's disease, although current research questions the efficiency of DBS for movement disorders.

Neuromodulation is a relatively new field that combines the use of neurodevices and neurochemistry. The basis of this field is that the brain can be regulated using a number of different factors (metabolic, electrical stimulation, physiological) and that all these can be modulated by devices implanted in the neural network. While currently this field is still in the researcher phase, it represents a new type of technological integration in the field of neurotechnology. The brain is a very sensitive organ, so in addition to researching the amazing things that neuromodulation and implanted neural devices can produce, it is important to research ways to create devices that elicit as few negative responses from the body as possible. This can be done by modifying the material surface chemistry of neural implants.

Cell Therapy

Researchers have begun looking at uses for stem cells in the brain, which recently have been found in a few loci. A large number of studies are being done to determine if this form of therapy could be used in a large scale. Experiments have successfully used stem cells in the brains of children who suffered from injuries in gestation and elderly people with degenerative diseases in order to induce the brain to produce new cells and to make more connections between neurons.

Pharmaceuticals

Pharmaceuticals play a vital role in maintaining stable brain chemistry, and are the most commonly used neurotechnology by the general public and medicine. Drugs like sertraline, methylphenidate, and zolpidem act as chemical modulators in the brain, and they allow for normal activity in many people whose brains cannot act normally under physiological conditions. While pharmaceuticals are usually not mentioned and have their own field, the role of pharmaceuticals is perhaps the most far-reaching and commonplace in modern society. Movement of magnetic particles to targeted brain regions for drug delivery is an emerging field of study and causes no detectable circuit damage.

Low Field Magnetic Stimulation

Stimulation with low-intensity magnetic fields is currently under study for depression at Harvard Medical School, and has previously been explored by Bell (et al.), Marino (et al.), and others.

How these Help Study the Brain

Magnetic resonance imaging is a vital tool in neurological research in showing activation in the brain as well as providing a comprehensive image of the brain being studied. While MRIs are used clinically for showing brain size, it still has relevance in the study of brains because it can be used to determine extent of injuries or deformation. These can have a significant effect on personality, sense perception, memory, higher order thinking, movement, and spatial understanding. However, current research tends to focus more so on fMRI or real-time functional MRI (rtfMRI). These two methods allow the scientist or the participant, respectively, to view activation in the brain. This is incredibly vital in understanding how a person thinks and how their brain reacts to a person's environment, as well as understanding how the brain works under various stressors or dysfunctions. Real-time functional MRI is a revolutionary tool available to neurologists and neuroscientists because patients can see how their brain reacts to stressors and can perceive visual feedback. CT scans are very similar to MRI in their academic use because they can be used to image the brain upon injury, but they are more limited in perceptual feedback. CTs are generally used in clinical studies far more than in academic studies, and are found far more often in a hospital than a research facility. PET scans are also finding more relevance in academia because they can be used to observe metabolic uptake of neurons, giving researchers a wider perspective about neural activity in the brain for a given condition. Combinations of these methods can provide researchers with knowledge of both physiological and metabolic behaviors of loci in the brain and can be used to explain activation and deactivation of parts of the brain under specific conditions.

Transcranial magnetic stimulation is a relatively new method of studying how the brain functions and is used in many research labs focused on behavioral disorders and hallucinations. What makes TMS research so interesting in the neuroscience community is that it can target specific regions of the brain and shut them down or activate temporarily; thereby changing the way the brain behaves. Personality disorders can stem from a variety of external factors, but when the disorder stems from the circuitry of the brain TMS can be used to deactivate the circuitry. This can give rise to a number of responses, ranging from "normality" to something more unexpected, but current research is based on the theory that use of TMS could radically change treatment and perhaps act as a cure for personality disorders and hallucinations. Currently, repetitive transcranial magnetic stimulation (rTMS) is being researched to see if this deactivation effect can be made more permanent in patients suffering from these disorders. Some techniques combine TMS and another scanning method such as EEG to get additional information about brain activity such as cortical response.

Both EEG and MEG are currently being used to study the brain's activity under different conditions. Each uses similar principles but allows researchers to examine individual regions of the brain, allowing isolation and potentially specific classification of active regions. As mentioned above, EEG is very useful in analysis of immobile patients, typically during the sleep cycle. While there are other types of research that utilize EEG, EEG has been fundamental in understanding the resting brain during sleep. There are other potential uses for EEG and MEG such as charting rehabilitation and improvement after trauma as well as testing neural conductivity in specific regions of epileptics or patients with personality disorders.

Neuromodulation can involve numerous technologies combined or used independently to achieve a desired effect in the brain. Gene and cell therapy are becoming more prevalent in research and clinical trials and these technologies could help stunt or even reverse disease progression in the central nervous system. Deep brain stimulation is currently used in many patients with movement disorders and is used to improve the quality of life in patients. While deep brain stimulation is a method to study how the brain functions per se, it provides both surgeons and neurologists important information about how the brain works when certain small regions of the basal ganglia (nuclei) are stimulated by electrical currents.

Future Technologies

The future of neurotechnologies lies in how they are fundamentally applied, and not so much on what new versions will be developed. Current technologies give a large amount of insight into the mind and how the brain functions, but basic research is still needed to demonstrate the more applied functions of these technologies. Currently, rtfMRI is being researched as a method for pain therapy. deCharms et al. have shown that there is a significant improvement in the way people perceive pain if they are made aware of how their brain is functioning while in pain. By providing direct and understandable feedback, researchers can help patients with chronic pain decrease their symptoms. This new type of bio/mechanical-feedback is a new development in pain therapy. Functional MRI is also being considered for a number of more applicable uses outside of the clinic. Research has been done on testing the efficiency of mapping the brain in the case when someone lies as a new way to detect lying. Along the same vein, EEG has been considered for use in lie detection as well. TMS is being used in a variety of potential therapies for patients with personality disorders, epilepsy, PTSD, migraine, and other brain-firing disorders, but has been found to have varying clinical success for each condition. The end result of such research would be to develop a method to alter the brain's perception and firing and train patients' brains to rewire permanently under inhibiting conditions. In addition, PET scans have been found to be 93% accurate in detecting Alzheimer's disease nearly 3 years before conventional diagnosis, indicating that PET scanning is becoming more useful in both the laboratory and the clinic.

Stem cell technologies are always salient both in the minds of the general public and scientists because of their large potential. Recent advances in stem cell research have allowed researchers to ethically pursue studies in nearly every facet of the body, which includes the brain. Research has shown that while most of the brain does not regenerate and is typically a very difficult environment to foster regeneration, there are portions of the brain with regenerative capabilities (specifically the hippocampus and the olfactory bulbs). Much of the research in central nervous system regeneration is how to overcome this poor regenerative quality of the brain. It is important to note that there are therapies that improve cognition and increase the amount of neural pathways, but this does not mean that there is a proliferation of neural cells in the brain. Rather, it is called a plastic rewiring of the brain (*plastic* because it indicates malleability) and is considered a vital part of growth. Nevertheless, many problems in patients stem from death of neurons in the brain, and researchers in the field are striving to produce technologies that enable regeneration in patients with stroke, Parkinson's diseases, severe trauma, and Alzheimer's disease, as well as many others. While still in fledgling stages of development, researchers have recently begun making very interesting progress in attempting to treat these diseases. Researchers have recently successfully produced dopaminergic neurons for transplant in patients with Parkinson's diseases with the hopes that they will be able to move again with a more steady supply of dopamine. Many researchers are building scaffolds that could be transplanted into a patient with spinal cord trauma to present an environment that promotes growth of axons (portions of the cell attributed with transmission of electrical signals) so that patients unable to move or feel might be able to do so again. The potentials are wide-ranging, but it is important to note that many of these therapies are still in the laboratory phase and are slowly being adapted in the clinic. Some scientists remain skeptical with the development of the field, and warn that there is a much larger chance that electrical prosthesis will be developed to solve clinical problems such as hearing loss or paralysis before cell therapy is used in a clinic.

Novel drug delivery systems are being researched in order to improve the lives of those who struggle with brain disorders that might not be treated with stem cells, modulation, or rehabilitation. Pharmaceuticals play a very important role in society, and the brain has a very selective barrier that prevents some drugs from going from the blood to the brain. There are some diseases of the brain such as meningitis that require doctors to directly inject medicine into the spinal cord because the drug cannot cross the blood–brain barrier. Research is being conducted to investigate new methods of targeting the brain using the blood supply, as it is much easier to inject into the blood than the spine. New technologies such as nanotechnology are being researched for selective drug delivery, but these technologies have problems as with any other. One of the major setbacks is that when a particle is too large, the patient's liver will take up the particle and degrade it for excretion, but if the particle is too small there will not be enough drug in the particle to take effect. In addition, the size of the capillary pore is important because too large a particle might not fit or even plug up the hole, preventing adequate supply of the drug to the brain. Other research is involved in integrating a protein device between the layers to create a free-flowing gate that is unimpeded by the limitations of the body. Another direction is receptor-mediated transport, where receptors in the brain used to transport nutrients are manipulated to transport drugs across the blood–brain barrier. Some have even suggested that focused ultrasound opens the blood–brain barrier momentarily and allows free passage of chemicals into the brain. Ultimately the goal for drug delivery is to develop a method that maximizes the amount of drug in the loci with as little degraded in the blood stream as possible.

Neuromodulation is a technology currently used for patients with movement disorders, although research is currently being done to apply this technology to other disorders. Recently, a study was done on if DBS could improve depression with positive results, indicating that this technology might have potential as a therapy for multiple disorders in the brain. DBS is limited by its high cost however, and in developing countries the availability of DBS is very limited. A new version of DBS is under investigation and has developed into the novel field, optogenetics. Optogenetics is the combination of deep brain stimulation with fiber optics and gene therapy. Essentially, the fiber optic cables are designed to light up under electrical stimulation, and a protein would be added to a neuron via gene therapy to excite it under light stimuli. So by combining these three independent fields, a surgeon could excite a single and specific neuron in order to help treat a patient with some disorder. Neuromodulation offers a wide degree of therapy for many patients, but due to the nature of the disorders it is currently used to treat its effects are often temporary. Future goals in the field hope to alleviate that problem by increasing the years of effect until DBS can be used for the remainder of the patient's life. Another use for neuromodulation would be in building neuro-interface prosthetic devices that would allow quadriplegics the ability to maneuver a cursor on a screen with their thoughts, thereby increasing their ability to interact with others around them. By understanding the motor cortex and understanding how the brain signals motion, it is possible to emulate this response on a computer screen.

Ethics

Stem Cells

The ethical debate about use of embryonic stem cells has stirred controversy both in the United States and abroad; although more recently these debates have lessened due to modern advances in creating induced pluripotent stem cells from adult cells. The greatest advantage for use of embryonic stem cells is the fact that they can differentiate (become) nearly any type of cell provided the right conditions and signals. However, recent advances by Shinya Yamanaka et al. have found ways to create pluripotent cells without the use of such controversial cell cultures. Using the patient's own cells and re-differentiating them into the desired cell type bypasses both possible patient rejection of the embryonic stem cells and any ethical concerns associated with using them, while also providing researchers a larger supply of available cells. However, induced pluripotent cells have the potential to form benign (though potentially malignant) tumors, and tend to have poor survivability *in vivo* (in the living body) on damaged tissue. Much of the ethics concerning use of stem cells has subsided from the embryonic/adult stem cell debate due to its rendered moot, but now societies find themselves debating whether or not this technology can be ethically used. Enhancements of traits, use of animals for tissue scaffolding, and even arguments for moral degeneration have been made with the fears that if this technology reaches its full potential a new paradigm shift will occur in human behavior.

Military Application

New neurotechnologies have always garnered the appeal of governments, from lie detection technology and virtual reality to rehabilitation and understanding the psyche. Due to the Iraq War and War on Terror, American soldiers coming back from Iraq and Afghanistan are reported to have percentages up to 12% with PTSD. There are many researchers hoping to improve these peoples' conditions by implementing new strategies for recovery. By combining pharmaceuticals and neurotechnologies, some researchers have discovered ways of lowering the "fear" response and theorize that it may be applicable to PTSD. Virtual reality is another technology that has drawn much attention in the military. If improved, it could be possible to train soldiers how to deal with complex situations in times of peace, in order to better prepare and train a modern army.

Privacy

Finally, when these technologies are being developed society must understand that these neurotechnologies could reveal the one thing that people can always keep secret: what they are thinking. While there are large amounts of benefits associated with these technologies, it is necessary for scientists and policy makers alike to consider implications about "cognitive liberty." This term is important in many ethical circles concerned with the state and goals of progress in the field of neurotechnology. Current improvements such as "brain fingerprinting" or lie detection using EEG or fMRI could give rise to a set fixture of loci/emotional relationships in the brain, although these technologies are still years away from full application. It is important to consider how all these neurotechnologies might affect the future of society, and it is suggested that political, scientific, and civil debates are heard about the implementation of these newer technologies that potentially offer a new wealth of once-private information. Some ethicists are also concerned with the use of TMS and fear that the technique could be used to alter patients in ways that are undesired by the patient.

Neural Oscillation

Neural oscillation is rhythmic or repetitive neural activity in the central nervous system. Neural tissue can generate oscillatory activity in many ways, driven either by mechanisms within individual neurons or by interactions between neurons. In individual neurons, oscillations can appear either as oscillations in membrane potential or as rhythmic patterns of action potentials, which then produce oscillatory activation of post-synaptic neurons. At the level of neural ensembles, synchronized activity of large numbers of neurons can give rise to macroscopic oscillations, which can be observed in an electroencephalogram. Oscillatory activity in groups of neurons generally arises from feedback connections between the neurons that result in the synchronization of their firing patterns. The interaction between neurons can give rise to oscillations at a different frequency than the firing frequency of individual neurons. A well-known example of macroscopic neural oscillations is alpha activity.

Neural oscillations were observed by researchers as early as 1924 (by Hans Berger). More than 50 years later, intrinsic oscillatory behavior was encountered in vertebrate neurons, but its functional role is still not fully understood. The possible roles of neural oscillations include feature binding, information transfer mechanisms and the generation of rhythmic motor output. Over the last decades more insight has been gained, especially with advances in brain imaging. A major area of research in neuroscience involves determining how oscillations are generated and what their roles are. Oscillatory activity in the brain is widely observed at different levels of observation and is thought to play a key role in processing neural information. Numerous experimental studies support a functional role of neural oscillations; a unified interpretation, however, is still lacking.

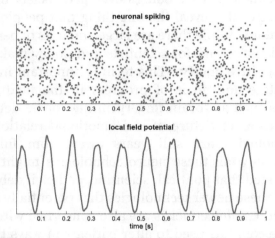

Simulation of neural oscillations at 10 Hz. Upper panel shows spiking of individual neurons (with each dot representing an individual action potential within the population of neurons), and the lower panel the local field potential reflecting their summed activity. Figure illustrates how synchronized patterns of action potentials may result in macroscopic oscillations that can be measured outside the scalp.

Overview

Neural oscillations are observed throughout the central nervous system at all levels, and include spike trains, local field potentials and large-scale oscillations which can be measured by electroencephalography (EEG). In general, oscillations can be characterized by their frequency, amplitude and phase. These signal properties can be extracted from neural recordings using time-frequency analysis. In large-scale oscillations, amplitude changes are considered to result from changes in synchronization within a neural ensemble, also referred to as local synchronization. In addition to local synchronization, oscillatory activity of distant neural structures (single neurons or neural ensembles) can synchronize. Neural oscillations and synchronization have been linked to many cognitive functions such as information transfer, perception, motor control and memory.

Neural oscillations have been most widely studied in neural activity generated by large groups of neurons. Large-scale activity can be measured by techniques such as EEG. In general, EEG signals have a broad spectral content similar to pink noise, but also reveal oscillatory activity in specific frequency bands. The first discovered and best-known frequency band is alpha activity (7.5–12.5 Hz) that can be detected from the occipital lobe during relaxed wakefulness and which increases when the eyes are closed. Other frequency bands are: delta (1–4 Hz), theta (4–8 Hz), beta (13–30 Hz) and gamma (30–70 Hz) frequency band, where faster rhythms such as gamma activity have been linked to cognitive processing. Indeed, EEG signals change dramatically during sleep and show a transition from faster frequencies to increasingly slower frequencies such as alpha waves. In fact, different sleep stages are commonly characterized by their spectral content. Consequently, neural oscillations have been linked to cognitive states, such as awareness and consciousness.

Although neural oscillations in human brain activity are mostly investigated using EEG recordings, they are also observed using more invasive recording techniques such as single-unit recordings. Neurons can generate rhythmic patterns of action potentials or spikes. Some types of neurons have the tendency to fire at particular frequencies, so-called *resonators*. Bursting is another form of rhythmic spiking. Spiking patterns are considered fundamental for information coding in the brain. Oscillatory activity can also be observed in the form of subthreshold membrane potential oscillations (i.e. in the absence of action potentials). If numerous neurons spike in synchrony, they can give rise to oscillations in local field potentials. Quantitative models can estimate the strength of neural oscillations in recorded data.

Neural oscillations are commonly studied from a mathematical framework and belong to the field of "neurodynamics", an area of research in the cognitive sciences that places a strong focus upon the dynamic character of neural activity in describing brain function. It considers the brain a dynamical system and uses differential equations to describe how neural activity evolves over time. In particular, it aims to relate dynamic patterns of brain activity to cognitive functions such as perception and memory. In very abstract form, neural oscillations can be analyzed analytically. When studied in a more physiologically realistic setting, oscillatory activity is generally studied using computer simulations of a computational model.

The functions of neural oscillations are wide ranging and vary for different types of oscillatory activity. Examples are the generation of rhythmic activity such as a heartbeat and the neural binding of sensory features in perception, such as the shape and color of an object. Neural oscillations also play an important role in many neurological disorders, such as excessive synchronization during seizure activity in epilepsy or tremor in patients with Parkinson's disease. Oscillatory activity can also be used to control external devices in brain–computer interfaces, in which subjects can control an external device by changing the amplitude of particular brain rhythmics.

Physiology

Oscillatory activity is observed throughout the central nervous system at all levels of organization. Three different levels have been widely recognized: the micro-scale (activity of a single neuron), the meso-scale (activity of a local group of neurons) and the macro-scale (activity of different brain regions).

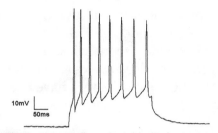

Tonic firing pattern of single neuron showing rhythmic spiking activity

Microscopic

Neurons generate action potentials resulting from changes in the electric membrane potential. Neurons can generate multiple action potentials in sequence forming so-called spike trains. These spike trains are the basis for neural coding and information transfer in the brain. Spike trains can form all kinds of patterns, such as rhythmic spiking and bursting, and often display oscillatory activity. Oscillatory activity in single neurons can also be observed in sub-threshold fluctuations in membrane potential. These rhythmic changes in membrane potential do not reach the critical threshold and therefore do not result in an action potential. They can result from postsynaptic potentials from synchronous inputs or from intrinsic properties of neurons.

Neuronal spiking can be classified by their activity patterns. The excitability of neurons can be subdivided in Class I and II. Class I neurons can generate action potentials with arbitrarily low frequency depending on the input strength, whereas Class II neurons generate action potentials in a certain frequency band, which is relatively insensitive to changes in input strength. Class II neurons are also more prone to display sub-threshold oscillations in membrane potential.

Mesoscopic

A group of neurons can also generate oscillatory activity. Through synaptic interactions the firing patterns of different neurons may become synchronized and the rhythmic changes in electric potential caused by their action potentials will add up (constructive interference). That is, synchronized firing patterns result in synchronized input into other cortical areas, which gives rise to large-amplitude oscillations of the local field potential. These large-scale oscillations can also be measured outside the scalp using electroencephalography (EEG) and magnetoencephalography (MEG). The electric potentials generated by single neurons are far too small to be picked up outside the scalp, and EEG or MEG activity always reflects the summation of the synchronous activity of thousands or millions of neurons that have similar spatial orientation. Neurons in a neural ensemble rarely all fire at exactly the same moment, i.e. fully synchronized. Instead, the probability of firing is rhythmically modulated such that neurons are more likely to fire at the same time, which gives rise to oscillations in their mean activity. As such, the frequency of large-scale oscillations does not need to match the firing pattern of individual neurons. Isolated cortical neurons fire regularly under certain conditions, but in the intact brain cortical cells are bombarded by highly fluctuating synaptic inputs and typically fire seemingly at random. However, if the probability of a large group of neurons is rhythmically modulated at a common frequency, they will generate oscillations in the mean field. Neural ensembles can generate oscillatory activity endogenously through local interactions between excitatory and inhibitory neurons. In particular, inhibitory interneurons play an important role in producing neural ensemble synchrony by generating a narrow window for effective excitation and rhythmically modulating the firing rate of excitatory neurons.

Macroscopic

Neural oscillation can also arise from interactions between different brain areas. Time delays play an important role here. Because all brain areas are bidirectionally coupled, these connections between brain areas form feedback loops. Positive feedback loops tends to cause oscillatory activity which frequency is inversely related to the delay time. An example of such a feedback loop is the connections between the thalamus and cortex. This thalamocortical network is able to generate oscillatory activity known as recurrent thalamo-cortical resonance. The thalamocortical network plays an important role in the generation of alpha activity.

Mechanisms

Neuronal Properties

Scientists have identified some intrinsic neuronal properties that play an important role in generating membrane potential oscillations. In particular, voltage-gated ion channels are critical in the generation of action potentials. The dynamics of these ion channels have been captured in the well-established Hodgkin–Huxley model that describes how action potentials are initiated and propagated by means of a set of differential equations. Using bifurcation analysis, different oscillatory varieties of these neuronal models can be determined, allowing for the classification of types of neuronal responses. The oscillatory dynamics of neuronal spiking as identified in the Hodgkin–Huxley model closely agree with empirical findings. In addition to periodic spiking, subthreshold membrane potential oscillations, i.e. resonance behavior that does not result in action potentials, may also contribute to oscillatory activity by facilitating synchronous activity of neighboring neurons. Like pacemaker neurons in central pattern generators, subtypes of cortical cells fire bursts of spikes (brief clusters of spikes) rhythmically at preferred frequencies. Bursting neurons have the potential to serve as pacemakers for synchronous network oscillations, and bursts of spikes may underlie or enhance neuronal resonance.

Network Properties

Apart from intrinsic properties of neurons, network properties are also an important source of oscillatory activity. Neurons communicate with one another via synapses and affect the timing of spike trains in the post-synaptic neurons. Depending on the properties of the connection, such as the coupling strength, time delay and whether coupling is excitatory or inhibitory, the spike trains of the interacting neurons may become synchronized. Neurons are locally connected, forming small clusters that are called neural ensembles. Certain network structures promote oscillatory activity at specific frequencies. For example, neuronal activity generated by two populations of interconnected *inhibitory* and *excitatory* cells can show spontaneous oscillations that are described by the Wilson-Cowan model.

If a group of neurons engages in synchronized oscillatory activity, the neural ensemble can be mathematically represented as a single oscillator. Different neural ensembles are coupled through long-range connections and form a network of weakly coupled oscillators at the next spatial scale. Weakly coupled oscillators can generate a range of dynamics including oscillatory activity. Long-range connections between different brain structures, such as the thalamus and the cortex, involve time-delays due to the finite conduction velocity of axons. Because most con-nections are reciprocal, they form feedback loops that support oscillatory activity. Oscillations recorded from multiple cortical areas can become synchronized to form large scale brain networks, whose dynamics and functional connectivity can be studied by means of spectral analysis and Granger causality measures. Coherent activity of large-scale brain activity may form dynamic links between brain areas required for the inte-gration of distributed information.

Neuromodulation

In addition to fast direct synaptic interactions between neurons forming a network, oscillatory activity is modulated by neurotransmitters on a much slower time scale. That is, the concentration levels of certain neurotransmitters are known to regulate the amount of oscillatory activity. For instance, GABA concentration has been shown to be positively correlated with frequency of oscillations in induced stimuli. A number of nuclei in the brainstem have diffuse projections throughout the brain influencing concentration levels of neurotransmitters such as norepinephrine, acetylcholine and serotonin. These neurotransmitter systems affect the physiological state, e.g., wakefulness or arousal, and have a pronounced effect on amplitude of different brain waves, such as alpha activity.

Mathematical Description

Oscillations can often be described and analyzed using mathematics. Mathematicians have identified several dynamical mechanisms that generate rhythmicity. Among the most important are harmonic (linear) oscillators, limit cycle oscillators, and de-layed-feedback oscillators. Harmonic oscillations appear very frequently in nature—examples are sound waves, the motion of a pendulum, and vibrations of every sort.

They generally arise when a physical system is perturbed by a small degree from a minimum-energy state, and are well-understood mathematically. Noise-driven harmonic oscillators realistically simulate alpha rhythm in the waking EEG as well as slow waves and spindles in the sleep EEG. Successful EEG analysis algorithms were based on such models. Several other EEG components are better described by limit-cycle or delayed-feedback oscillations. Limit-cycle oscillations arise from physical systems that show large deviations from equilibrium, whereas delayed-feedback oscillations arise when components of a system affect each other after significant time delays. Limit-cycle oscillations can be complex but there are powerful mathematical tools for analyzing them; the mathematics of delayed-feedback oscillations is primitive in comparison. Linear oscillators and limit-cycle oscillators qualitatively differ in terms of how they respond to fluctuations in input. In a linear oscillator, the frequency is more or less constant but the amplitude can vary greatly. In a limit-cycle oscillator, the amplitude tends to be more or less constant but the frequency can vary greatly. A heartbeat is an example of a limit-cycle oscillation in that the frequency of beats varies widely, while each individual beat continues to pump about the same amount of blood.

Computational models adopt a variety of abstractions in order to describe complex oscillatory dynamics observed in brain activity. Many models are used in the field, each defined at a different level of abstraction and trying to model different aspects of neural systems. They range from models of the short-term behaviour of individual neurons, through models of how the dynamics of neural circuitry arise from interactions between individual neurons, to models of how behaviour can arise from abstract neural modules that represent complete subsystems.

Single Neuron Model

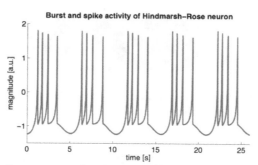

Simulation of a Hindmarsh–Rose neuron showing typical bursting behavior: a fast rhythm generated by individual spikes and a slower rhythm generated by the bursts.

A model of a biological neuron is a mathematical description of the properties of nerve cells, or neurons, that is designed to accurately describe and predict its biological processes. The most successful and widely used model of neurons, the Hodgkin–Huxley model, is based on data from the squid giant axon. It is a set of nonlinear ordinary differential equations that approximates the electrical characteristics of a neuron, in particular the generation and propagation of action potentials. The model is very accurate and detailed and Hodgkin and Huxley received the 1963 Nobel Prize in physiology or medicine for this work.

The mathematics of the Hodgkin–Huxley model are quite complicated and several simplifications have been proposed, such as the FitzHugh–Nagumo model and the Hindmarsh–Rose model. Such models only capture the basic neuronal dynamics, such as rhythmic spiking and bursting, but are more computationally efficient. This allows the simulation of a large number of interconnected neurons that form a neural network.

Spiking Model

A neural network model describes a population of physically interconnected neurons or a group of disparate neurons whose inputs or signalling targets define a recognizable circuit. These models aim to describe how the dynamics of neural circuitry arise from interactions between individual neurons. Local interactions between neurons can result in the synchronization of spiking activity and form the basis of oscillatory activity. In particular, models of interacting pyramidal cells and inhibitory interneurons have been shown to generate brain rhythms such as gamma activity.

Neural Mass Model

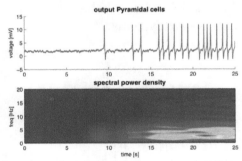

Simulation of a neural mass model showing network spiking during the onset of a seizure. As the gain A is increased the network starts to oscillate at 3Hz.

Neural field models are another important tool in studying neural oscillations and are a mathematical framework describing evolution of variables such as mean firing rate in space and time. In modeling the activity of large numbers of neurons, the central idea is to take the density of neurons to the continuum limit, resulting in spatially continuous neural networks. Instead of modelling individual neurons, this approach approximates a group of neurons by its average properties and interactions. It is based on the mean field approach, an area of statistical physics that deals with large-scale systems. Models based on these principles have been used to provide mathematical descriptions of neural oscillations and EEG rhythms. They have for instance been used to investigate visual hallucinations.

Kuramoto Model

The Kuramoto model of coupled phase oscillators is one of the most abstract and fundamental model used to investigate neural oscillations and synchronization. It captures the activity of a local system (e.g., a single neuron or neural ensemble) by its circular phase alone and hence ignores the amplitude of oscillations (amplitude is constant). Interactions amongst these oscillators are introduced by a simple algebraic form (such as a sine function) and collectively generate a dynamical pattern at the global scale. The Kuramoto model is widely used to study oscillatory brain activity and several extensions have been proposed that increase its neurobiological plausibility, for instance by incorporating topological properties of local cortical connectivity. In particular, it describes how the activity of a group of interactioning neurons can become synchronized and generate large-scale oscillations. Simulations using the Kuramoto model with realistic long-range cortical connectivity and time-delayed interactions reveal the emergence of slow patterned fluctuations that reproduce resting-state BOLD functional maps, which can be measured using fMRI.

Activity Patterns

Both single neurons and groups of neurons can generate oscillatory activity spontaneously. In addition, they may show oscillatory responses to perceptual input or motor output. Some types of neurons will fire rhythmically in the absence of any synaptic input. Likewise, brain-wide activity reveals oscillatory activity while subjects do not engage in any activity, so-called resting-state activity. These ongoing rhythms can change in different ways in response to perceptual input or motor output. Oscillatory activity may respond by increases or decreases in frequency and amplitude or show a temporary interruption, which is referred to as phase resetting. In addition, external activity may not interact with ongoing activity at all, resulting in an additive response.

Ongoing Activity

Spontaneous activity is brain activity in the absence of an explicit task, such as sensory input or motor output, and hence also referred to as resting-state activity. It is opposed to induced activity, i.e. brain activity that is induced by sensory stimuli or motor responses. The term *ongoing brain activity* is used in electroencephalography and magnetoencephalography for those signal components that are not associated with the processing of a stimulus or the occurrence of specific other events, such as moving a body part, i.e. events that do not form evoked potentials/evoked fields, or induced activity. Spontaneous activity is usually considered to be noise if one is interested in stimulus processing; however, spontaneous activity is considered to play a crucial role during brain development, such as in network formation and synaptogenesis. Spontaneous activity may be informative regarding the current mental state of the person (e.g. wakefulness, alertness) and is often used in sleep research. Certain types of oscillatory activity, such as alpha waves, are part of spontaneous activity. Statistical analysis of power fluctuations of alpha activity reveals a bimodal distribution, i.e. a high- and low-amplitude mode, and hence shows that resting-state activity does not just reflect a noise process. In case of fMRI, spontaneous fluctuations in the Blood-oxygen-level dependent (BOLD) signal reveal correlation patterns that are linked to resting states networks, such as the default network. The temporal evolution of resting state networks is correlated with fluctuations of oscillatory EEG activity in different frequency bands.

Ongoing brain activity may also have an important role in perception, as it may interact with activity related to incoming stimuli. Indeed, EEG studies suggest that visual perception is dependent on both the phase and amplitude of cortical oscillations. For instance, the amplitude and phase of alpha activity at the moment of visual stimulation predicts whether a weak stimulus will be perceived by the subject.

Frequency Response

In response to input, a neuron or neuronal ensemble may change the frequency at which it oscillates, thus changing the rate at which it spikes. Often, a neuron's firing rate depends on the summed activity it receives. Frequency changes are also commonly observed in central pattern generators and directly relate to the speed of motor activities, such as step frequency in walking. However, changes in *relative* oscillation frequency between different brain areas is not so common because the frequency of oscillatory activity is often related to the time delays between brain areas.

Phase Resetting

Phase resetting occurs when input to a neuron or neuronal ensemble resets the phase of ongoing oscillations. It is very common in single neurons where spike timing is adjusted to neuronal input (a neuron may spike at a fixed delay in response to periodic input, which is referred to as phase locking) and may also occur in neuronal ensembles when the phases of their neurons are adjusted simultaneously. Phase resetting is fundamental for the synchronization of different neurons or different brain regions because the timing of spikes can become phase locked to the activity of other neurons.

Phase resetting also permits the study of evoked activity, a term used in electroencephalography and magnetoencephalography for responses in brain activity that are directly related to stimulus-related activity. Evoked potentials and event-related potentials are obtained from an electroencephalogram by stimulus-locked averaging, i.e. averaging different trials at fixed latencies around the presentation of a stimulus. As a consequence, those signal components that are the same in each single measurement are conserved and all others, i.e. ongoing or spontaneous activity, are averaged out. That is, event-related potentials only reflect oscillations in brain activity that are phase-locked to the stimulus or event. Evoked activity is often considered to be independent from ongoing brain activity, although this is an ongoing debate.

Amplitude Response

Next to evoked activity, neural activity related to stimulus processing may result in induced activity. Induced activity refers to modulation in ongoing brain activity induced by processing of stimuli or movement preparation. Hence, they reflect an indirect response in contrast to evoked responses. A well-studied type of induced activity is amplitude change in oscillatory activity. For instance, gamma activity often increases during increased mental activity such as during object representation. Because induced responses may have different phases across measurements and therefore would cancel out during averaging, they can only be obtained using time-frequency analysis. Induced activity generally reflects the activity of numerous neurons: amplitude changes in oscillatory activity are thought to arise from the synchronization of neural activity, for instance by synchronization of spike timing or membrane potential fluctuations of individual neurons. Increases in oscillatory activity are therefore often referred to as event-related synchronization, while decreases are referred to as event-related desynchronization.

Asymmetric Amplitude Modulation

It has recently been proposed that even if phases are not aligned across trials, induced activity may still cause event-related potentials because ongoing brain oscillations may not be symmetric and thus amplitude modulations may result in a baseline shift that does not average out. This model implies that slow event-related responses, such as asymmetric alpha activity, could result from asymmetric brain oscillation amplitude modulations, such as an asymmetry of the intracellular currents that propagate forward and backward down the dendrites. Under this assumption, asymmetries in the dendritic current would cause asymmetries in oscillatory activity measured by EEG and MEG, since dendritic currents in pyramidal cells are generally thought to generate EEG and MEG signals that can be measured at the scalp.

Function

Neural synchronization can be modulated by task constraints, such as attention, and is thought to play a role in feature binding, neuronal communication, and motor coordination. Neuronal oscillations became a hot topic in neuroscience in the 1990s when the studies of the visual system of the brain by Gray, Singer and others appeared to support the neural binding hypothesis. According to this idea, synchronous oscillations in neuronal ensembles bind neurons representing different features of an object. For example, when a person looks at a tree, visual cortex neurons representing the tree trunk and those representing the branches of the same tree would oscillate in synchrony to form a single representation of the tree. This phenomenon is best seen in local field potentials which reflect the synchronous activity of local groups of neurons, but has also been shown in EEG and MEG recordings providing increasing evidence for a close relation between synchronous oscillatory activity and a variety of cognitive functions such as perceptual grouping.

Pacemaker

Cells in the sinoatrial node, located in the right atrium of the heart, spontaneously depolarize approximately 100 times per minute. Although all of the heart's cells have the ability to generate action potentials that trigger cardiac contraction, the sinoatrial node normally initiates it, simply because it generates impulses slightly faster than the other areas. Hence, these cells generate the normal sinus rhythm and are called pacemaker cells as they directly control the heart rate. In the absence of extrinsic neural and hormonal control, cells in the SA node will rhythmically discharge. The sinoatrial node is richly innervated by the autonomic nervous system, which up or down regulates the spontaneous firing frequency of the pacemaker cells.

Central Pattern Generator

Synchronized firing of neurons also forms the basis of periodic motor commands for rhythmic movements. These rhythmic outputs are produced by a group of interacting neurons that form a network, called a central pattern generator. Central pattern generators are neuronal circuits that - when activated - can produce rhythmic motor patterns in the absence of sensory or descending inputs that carry specific timing information. Examples are walking, breathing, and swimming, Most evidence for central pattern generators comes from lower animals, such as the lamprey, but there is also evidence for spinal central pattern generators in humans.

Information Processing

Neuronal spiking is generally considered the basis for information transfer in the brain. For such a transfer, information needs to be coded in a spiking pattern. Different types of coding schemes have been proposed, such as rate coding and temporal coding.

Perception

Synchronization of neuronal firing may serve as a means to group spatially segregated neurons that respond to the same stimulus in order to bind these responses for further joint processing, i.e. to exploit temporal synchrony to encode relations. Purely theoretical formulations of the binding-by-synchrony hypothesis were proposed first, but subsequently extensive experimental evidence has been reported supporting the potential role of synchrony as a relational code.

The functional role of synchronized oscillatory activity in the brain was mainly established in experiments performed on awake kittens with multiple electrodes implanted in the visual cortex. These experiments showed that groups of spatially segregated neurons engage in synchronous oscillatory activity when activated by visual stimuli. The frequency of these oscillations was in the range of 40 Hz and differed from the periodic activation induced by the grating, suggesting that the oscillations and their synchronization were due to internal neuronal interactions. Similar findings were shown in parallel by the group of Eckhorn, providing further evidence for the functional role of neural synchronization in feature binding. Since then, numerous studies have replicated these findings and extended them to different modalities such as EEG, providing extensive evidence of the functional role of gamma oscillations in visual perception.

Gilles Laurent and colleagues showed that oscillatory synchronization has an important functional role in odor perception. Perceiving different odors leads to different subsets of neurons firing on different sets of oscillatory cycles. These oscillations can be disrupted by GABA blocker picrotoxin, and the disruption of the oscillatory synchronization leads to impairment of behavioral discrimination of chemically similar odorants in bees and to more similar responses across odors in downstream β-lobe neurons.

Neural oscillations are also thought be involved in the sense of time and in somatosensory perception. However, recent findings argue against a clock-like function of cortical gamma oscillations.

Motor Coordination

Oscillations have been commonly reported in the motor system. Pfurtscheller and colleagues found a reduction in alpha (8–12 Hz) and beta (13–30 Hz) oscillations in EEG activity when subjects made a movement. Using intra-cortical recordings, similar changes in oscillatory activity were found in the motor cortex when the monkeys performed motor acts that required significant attention. In addition, oscillations at spinal level become synchronised to beta oscillations in the motor cortex during constant muscle activation, as determined by cortico-muscular coherence. Likewise, muscle activity of different muscles reveals inter-muscular coherence at multiple distinct frequencies reflecting the underlying neural circuitry involved in motor coordination.

Recently it was found that cortical oscillations propagate as travelling waves across the surface of the motor cortex along dominant spatial axes characteristic of the local circuitry of the motor cortex. It has been proposed that motor commands in the form of travelling waves can be spatially filtered by the descending fibres to selectively control muscle force. Simulations have shown that ongoing wave activity in cortex can elicit steady muscle force with physiological levels of EEG-EMG coherence.

Oscillatory rhythms at 10 Hz have been recorded in a brain area called the inferior olive, which is associated with the cerebellum. These oscillations are also observed in motor output of physiological tremor and when performing slow finger movements. These findings may indicate that the human brain controls continuous movements intermittently. In support, it was shown that these movement discontinuities are directly correlated to oscillatory activity in a cerebello-thalamo-cortical loop, which may represent a neural mechanism for the intermittent motor control.

Memory

Neural oscillations, in particular theta activity, are extensively linked to memory function. Theta rhythms are very strong in rodent hippocampi and entorhinal cortex during learning and memory retrieval, and they are believed to be vital to the induction of long-term potentiation, a potential cellular mechanism for learning and memory. Coupling between theta and gamma activity is thought to be vital for memory functions, including episodic memory. Tight coordination of single-neuron spikes with local theta oscillations is linked to successful memory formation in humans, as more stereotyped spiking predicts better memory.

Sleep and Consciousness

Sleep is a naturally recurring state characterized by reduced or absent consciousness and proceeds in cycles of rapid eye movement (REM) and non-rapid eye movement (NREM) sleep. The normal order of sleep stages is N1 → N2 → N3 → N2 → REM. Sleep stages are characterized by spectral content of EEG: for instance, stage N1 refers to the transition of the brain from alpha waves (common in the awake state) to theta waves, whereas stage N3 (deep or slow-wave sleep) is characterized by the presence of delta waves.

Development

Neural oscillations may play a role in neural development. For example, retinal waves are thought to have properties that define early connectivity of circuits and synapses between cells in the retina.

Pathology

Catherine Metzger

13 Octobre 1869

Handwriting of a person affected by Parkinson's disease showing rhythmic tremor activity in the strokes

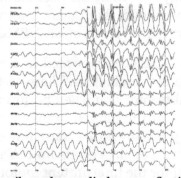

Generalized 3 Hz spike and wave discharges reflecting seizure activity

Specific types of neural oscillations may also appear in pathological situations, such as Parkinson's disease or epilepsy. Interestingly, these pathological oscillations often consist of an aberrant version of a normal oscillation. For example, one of the best known types is the spike and wave oscillation, which is typical of generalized or absence epileptic seizures, and which resembles normal sleep spindle oscillations.

Tremor

A tremor is an involuntary, somewhat rhythmic, muscle contraction and relaxation involving to-and-fro movements of one or more body parts. It is the most common of all involuntary movements and can affect the hands, arms, eyes, face, head, vocal cords, trunk, and legs. Most tremors occur in the hands. In some people, tremor is a symptom of another neurological disorder. Many different forms of tremor have been identified, such as essential tremor or Parkinsonian tremor. It is argued that tremors are likely to be multifactorial in origin, with contributions from neural oscillations in the central nervous systems, but also from peripheral mechanisms such as reflex loop resonances.

Epilepsy

Epilepsy is a common chronic neurological disorder characterized by seizures. These seizures are transient signs and/or symptoms of abnormal, excessive or hypersynchronous neuronal activity in the brain.

Applications

Clinical Endpoints

Neural oscillations are sensitive to several drugs influencing brain activity; accordingly, biomarkers based on neural oscillations are emerging as secondary endpoints in clinical trials and in quantifying effects in pre-clinical studies. Theses biomarkers are often named "EEG biomarkers" or "Neurophysiological Biomarkers" and are quantified using Quantitative electroencephalography (qEEG). EEG biomarkers can be extracted from the EEG using the open-source Neurophysiological Biomarker Toolbox.

Brain–computer Interface

Neural oscillation has been applied as a control signal in various brain–computer interfaces (BCIs). For example, a non-invasive BCI interface can be created by placing electrodes on the scalp and then measuring the weak electric signals. Although individual neuron activities cannot be recovered through non-invasive BCI because the skull damps and blurs the electromagnetic signals, oscillatory activity can still be reliably detected. In particular, some forms of BCI allow users to control a device by measuring the amplitude of oscillatory activity in specific frequency bands, including mu and beta rhythms.

Examples

A non-inclusive list of types of oscillatory activity found in the central nervous system:

- Delta wave
- Theta wave
- Alpha wave
- Mu wave
- Beta wave
- Gamma wave
- Sleep spindle
- Thalamocortical oscillations
- Subthreshold membrane potential oscillations
- Bursting
- Cardiac cycle
- Epileptic seizure
- Mathematical modeling of electrophysiological activity in epilepsy
- Sharp wave–ripple complexes

Neural Backpropagation

Neural backpropagation is the phenomenon in which the action potential of a neuron creates a voltage spike both at the end of the axon (normal propagation) and back through to the dendritic arbor or dendrites, from which much of the original input current originated. It has been shown that this simple process can be used in a manner similar to the backpropagation algorithm used in multilayer perceptrons, a type of artificial neural network. In addition to active backpropagation of the action potential, there is also passive electrotonic spread. While there is ample evidence to prove the existence of backpropagating action potentials, the function of such action potentials and the extent to which they invade the most distal dendrites remains highly controversial.

Mechanism

When a neuron fires an action potential, it is initiated at the axon initial segment. An action potential spreads down the axon because of the gating properties of voltage-gated sodium channels and voltage-gated potassium channels. Initially, it was thought that an action potential could only travel down the axon in one direction towards the axon terminal where it ultimately signaled the release of neurotransmitters. However, recent research has provided evidence for the existence of backwards propagating action potentials (Staley 2004).

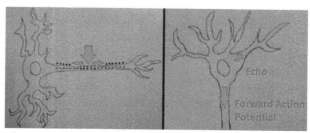

Methods of neural backpropagation. Left: action potential forms in axon and travels towards soma. Right: Regular action potential generates an echo that backpropagates through the dendritic tree.

Neural backpropagation can occur in one of two ways. First, during the initiation of an axonal action potential, the cell body, or soma, can become depolarized as well. This depolarization can spread through the cell body towards the dendritic tree where there are voltage-gated sodium channels. The depolarization of these voltage-gated sodium channels can then result in the propagation of a dendritic action potential. Such backpropagation is sometimes referred to as an echo of the forward propagating action potential (Staley 2004). It has also been shown that an action potential initiated in the axon can create a retrograde signal that travels in the opposite direction (Hausser 2000). This impulse travels up the axon eventually causing the cell body to become depolarized, thus triggering the dendritic voltage-gated calcium channels. As described in the first process, the triggering of dendritic voltage-gated calcium channels leads to the propagation of a dendritic action potential.

Generally, EPSPs from synaptic activation are not large enough to activate the dendritic voltage-gated calcium channels (usually on the order of a couple milliamperes each) so backpropagation is typically believed to happen only when the cell is activated to fire an action potential.

It is important to note that the strength of backpropagating action potentials varies greatly between different neuronal types (Hausser 2000). Some types of neuronal cells show little to no decrease in the amplitude of action potentials as they invade and travel through the dendritic tree while other neuronal cell types, such as cerebellar Purkinje neurons, exhibit very little action potential backpropagation (Stuart 1997). Additionally, there are other neuronal cell types that manifest varying degrees of amplitude decrement during backpropagation. It is thought that this is due to the fact that each neuronal cell type contains varying numbers of the voltage-gated channels required to propagate a dendritic action potential.

Regulation and Inhibition

Generally, synaptic signals that are received by the dendrite are combined in the soma in order to generate an action potential that is then transmitted down the axon toward the next synaptic contact. Thus, the backpropagation of action potentials poses a threat to initiate an uncontrolled positive-feedback loop between the soma and the dendrites. For example, as an action potential was triggered, its dendritic echo could enter the dendrite and potentially trigger a second action potential. If left unchecked, an endless cycle of action potentials triggered by their own echo would be created. In order to prevent such a cycle, most neurons have a relatively high density of A-type K+ channels.

A-type K+ channels belong to the superfamily of voltage-gated ion channels and are transmembrane channels that help maintain the cell's membrane potential (Cai 2007). Typically, they play a crucial role in returning the cell to its resting membrane following an action potential by allowing an inhibitory current of K+ ions to quickly flow out of the neuron. The presence of these channels in such high density in the dendrites explains their inability to initiate an action potential, even during synaptic input. Additionally, the presence of these channels provides a mechanism by which the neuron can suppress and regulate the backpropagation of action potentials through the dendrite (Vetter 2000). Results have indicated a linear increase in the density of A-type channels with increasing distance into the dendrite away from the soma. The increase in the density of A-type channels results in a dampening of the backpropagating action potential as it travels into the dendrite. Essentially, inhibition occurs because the A-type channels facilitate the outflow of K+ ions in order to maintain the membrane potential below threshold levels (Cai 2007). Such inhibition limits EPSP and protects the neuron from entering a never-ending positive-positive feedback loop between the soma and the dendrites.

History

Since the 1950s, evidence has existed that neurons in the central nervous system generate an action potential, or voltage spike, that travels both through the axon to signal the next neuron and backpropagates through the dendrites sending a retrograde signal to its presynaptic signaling neurons. This current decays significantly with travel length along the dendrites, so effects are predicted to be more significant for neurons whose synapses are near the postsynaptic cell body, with magnitude depending mainly on sodium-channel density in the dendrite. It is also dependent on the shape of the dendritic tree and, more importantly, on the rate of signal currents to the neuron. On average, a backpropagating spike loses about half its voltage after traveling nearly 500 micrometres.

Backpropagation occurs actively in the neocortex, hippocampus, substantia nigra, and spinal cord, while in the cerebellum it occurs relatively passively. This is consistent with observations that synaptic plasticity is much more apparent in areas like the hippocampus, which controls spatial memory, than the cerebellum, which controls more unconscious and vegetative functions.

The backpropagating current also causes a voltage change that increases the concentration of Ca^{2+} in the dendrites, an event which coincides with certain models of synaptic plasticity. This change also affects future integration of signals, leading to at least a short-term response difference between the presynaptic signals and the postsynaptic spike.

Functions

While many questions have yet to be answered in regards to neural backpropagation, there exists a number of hypotheses regarding its function. Some proposed function include involvement in synaptic plasticity, involvement in dendrodendritic inhibition, boosting synaptic responses, resetting membrane potential, retrograde actions at synapses and conditional axonal output. Backpropagation is believed to help form LTP (long term potentiation) and Hebbian plasticity at hippocampal synapses. Since artificial LTP induction, using microelectrode stimulation, voltage clamp, etc. requires the postsynaptic cell to be slightly depolarized when EPSPs are elicited, backpropagation can serve as the means of depolarization of the postsynaptic cell.

Algorithm

While a backpropagating action potential can presumably cause changes in the weight of the presynaptic connections, there is no simple mechanism for an error signal to propagate through multiple layers of neurons, as in the computer backpropagation algorithm. However, simple linear topologies have shown that effective computation is possible through signal backpropagation in this biological sense.

References

- Luger, George; Stubblefield, William (2004). Artificial Intelligence: Structures and Strategies for Complex Problem Solving (5th ed.). Benjamin/Cummings. ISBN 0-8053-4780-1.

- Neapolitan, Richard; Jiang, Xia (2012). Contemporary Artificial Intelligence. Chapman & Hall/CRC. ISBN 978-1-4398-4469-4.

- Norvig, Peter (2003), Artificial Intelligence: A Modern Approach (2nd ed.), Upper Saddle River, New Jersey: Prentice Hall, ISBN 0-13-790395-2 .

- Russell, Stuart J.; Norvig, Peter (2009), Artificial Intelligence: A Modern Approach (3rd ed.), Upper Saddle River, New Jersey: Prentice Hall, ISBN 0-13-604259-7 .

- Poole, David; Mackworth, Alan; Goebel, Randy (1998). Computational Intelligence: A Logical Approach. New York: Oxford University Press. ISBN 0-19-510270-3.

- Winston, Patrick Henry (1984). Artificial Intelligence. Reading, MA: Addison-Wesley. ISBN 0-201-08259-4.

- Bundy, Alan (1980). Artificial Intelligence: An Introductory Course (2nd ed.). Edinburgh University Press. ISBN 0-85224-410-X.

- Crevier, Daniel (1993), AI: The Tumultuous Search for Artificial Intelligence, New York, NY: BasicBooks, ISBN 0-465-02997-3 .

- McCorduck, Pamela (2004), Machines Who Think (2nd ed.), Natick, MA: A. K. Peters, Ltd., ISBN 1-56881-205-1 .

- Newquist, HP (1994). The Brain Makers: Genius, Ego, And Greed In The Quest For Machines That Think. New York: Macmillan/SAMS. ISBN 0-672-30412-0.

- Nilsson, Nils (2009). The Quest for Artificial Intelligence: A History of Ideas and Achievements. New York: Cambridge University Press. ISBN 978-0-521-12293-1.

- Dreyfus, Hubert; Dreyfus, Stuart (1986). Mind over Machine: The Power of Human Intuition and Expertise in the Era of the Computer. Oxford, UK: Blackwell. ISBN 0-02-908060-6.

- Fearn, Nicholas (2007). The Latest Answers to the Oldest Questions: A Philosophical Adventure with the World's Greatest Thinkers. New York: Grove Press. ISBN 0-8021-1839-9.

- Hofstadter, Douglas (1979). Gödel, Escher, Bach: an Eternal Golden Braid. New York, NY: Vintage Books. ISBN 0-394-74502-7.

- Kahneman, Daniel; Slovic, D.; Tversky, Amos (1982). Judgment under uncertainty: Heuristics and biases. New York: Cambridge University Press. ISBN 0-521-28414-7.

- Koza, John R. (1992). Genetic Programming (On the Programming of Computers by Means of Natural Selection). MIT Press. ISBN 0-262-11170-5.

- Lakoff, George; Núñez, Rafael E. (2000). Where Mathematics Comes From: How the Embodied Mind Brings Mathematics into Being. Basic Books. ISBN 0-465-03771-2.

- Minsky, Marvin (1967). Computation: Finite and Infinite Machines. Englewood Cliffs, N.J.: Prentice-Hall. ISBN 0-13-165449-7.

- O'Brien, James; Marakas, George (2011). Management Information Systems (10th ed.). McGraw-Hill/Irwin. ISBN 978-0-07-337681-3.

- Penrose, Roger (1989). The Emperor's New Mind: Concerning Computer, Minds and The Laws of Physics. Oxford University Press. ISBN 0-19-851973-7.

- Shapiro, Stuart C. (1992). "Artificial Intelligence". In Shapiro, Stuart C. Encyclopedia of Artificial Intelligence (PDF) (2nd ed.). New York: John Wiley. pp. 54–57. ISBN 0-471-50306-1.

- Weizenbaum, Joseph (1976). Computer Power and Human Reason. San Francisco: W.H. Freeman & Company. ISBN 0-7167-0464-1.

Theories and Models in Computational Neuroscience

There are several mathematical models used to explain the electrical activity of neurons. This chapter explores the main theories and models like biological neuron model, efficient coding hypothesis, Hodgkin–Huxley model, binding neuron etc. The chapter provides elaborate and insightful information about each model, its main features, objectives and applications.

Biological Neuron Model

A biological neuron model, also known as a spiking neuron model, is a mathematical model of the electrical properties of neuronal action potentials, which are sharp changes in the electrical potential across the cell membrane that last for about one millisecond. Spiking neurons are known to be a major signaling unit of the nervous system, and for this reason characterizing their operation is of great importance. It is worth noting that not all the cells of the nervous system produce the type of spike that define the scope of the spiking neuron models. For example, cochlear hair cells, retinal receptor cells, and retinal bipolar cells do not spike. Furthermore, many cells in the nervous system are not classified as neurons but instead are classified as glia.

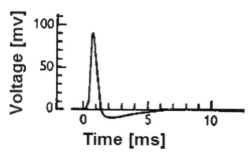

Fig. A neuronal action potential ("spike"). Note that the amplitude and the exact shape of the action potential can vary according to the exact experimental technique used for acquiring the signal.

Ultimately, biological neuron models aim to explain the mechanisms underlying the operation of the nervous system for the purpose of restoring lost control capabilities such as perception (e.g. deafness or blindness), motor movement decision making, and continuous limb control. In that sense, biological neural models differ from artificial neuron models that do not presume to predict the outcomes of experiments involving the biological neural tissue (although artificial neuron models are also concerned with execution of perception and estimation tasks). Accordingly, an important aspect of biological neuron models is experimental validation, and the use of physical units to describe the experimental procedure associated with the model predictions.

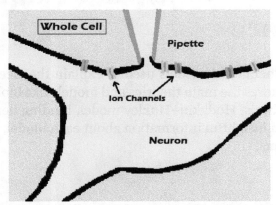

Fig. "Whole cell" measurement technique, which captures the spiking activity of a single neuron and produces full amplitude action potentials.

Neuron models can be divided into two categories according to the physical units of the interface of the model. Each category could be further divided according to the abstraction/detail level:

1. Electrical input–output membrane voltage models - These models produce a prediction for membrane output voltage as function of electrical stimulation at the input stage (either voltage or current). The various models in this category differ in the exact functional relationship between the input current and the output voltage and in the level of details. Some models in this category are black box models and distinguish only between two measured voltage levels: the presence of a spike (also known as "action potential") or a quiescent state. Other models are more detailed and account for sub-cellular processes.

2. Natural or pharmacological input neuron models - These models were inspired from experiments involving either natural or pharmacological stimulation. The results of these experiment tend to vary from trial to trial, but the averaged response tends to converge to a clear pattern. Accordingly, the output of natural and pharmacological neuron models is the probability of a spike event as function of the input stimulus. Typically, the output probability is normalized (divided by) a time constant, and the resulting normalized probability is called the "firing rate" and has units of Hertz. The models in this category differ in the functional relationship connecting the input stimulus to the output probability. Models that are sub-categorized as Markov models are the simplest and yield the most tractable results.

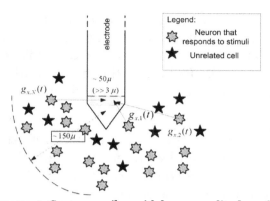

Fig. Extracellular measurement: Captures spikes with lower amplitudes, often from several spiking sources, depending on the size of the electrode and its proximity to the sources. Despite the decreased amplitude levels produced by this technique, it also has several advantages: 1) Is easier to obtain experimentally; 2) Is robust and lasts for a longer time; 3) Can reflect the dominant effect, especially when conducted in an anatomical region with many similar cells.

Although it is not unusual in science and engineering to have several descriptive models for different abstraction/detail levels, the number of different, sometimes contradicting, biological neuron models is exceptionally high. This situation is partly the result of the many different experimental settings, and the difficulty to separate the intrinsic properties of a single neuron from measurements effects and interactions of many cells (network effects). To accelerate the convergence to a unified theory, we list several models in each category, and where applicable, also references to supporting experiments.

Electrical Input–output Membrane Voltage Models

The models in this category describe the relationship between neuronal membrane currents at the input stage, and membrane voltage at the output stage. The most extensive experimental inquiry in this category of models was made by Hodgkin–Huxley in the early 1950s using an experimental setup that punctured the cell membrane and allowed to force a specific membrane voltage/current.

Most modern electrical neural interfaces apply extra-cellular electrical stimulation to avoid membrane puncturing which can lead to cell death and tissue damage. Hence, it is not clear to what extent the electrical neuron models hold for extra-cellular stimulation.

Integrate-and-fire

One of the earliest models of a neuron was first investigated in 1907 by Louis Lapicque. A neuron is represented in time by

$$I(t) = C_{\mathrm{m}} \frac{dV_{\mathrm{m}}(t)}{dt}$$

which is just the time derivative of the law of capacitance, $Q = CV$. When an input current is applied, the membrane voltage increases with time until it reaches a constant threshold V_{th}, at which point a delta function spike occurs and the voltage is reset to its resting potential, after which the model continues to run. The *firing frequency* of the model thus increases linearly without bound as input current increases.

The model can be made more accurate by introducing a refractory period t_{ref} that limits the firing frequency of a neuron by preventing it from firing during that period. Through some calculus involving a Fourier transform, the firing frequency as a function of a constant input current thus looks like

$$f(I) = \frac{I}{C_m V_{th} + t_{ref} I}.$$

A remaining shortcoming of this model is that it implements no time-dependent memory. If the model receives a below-threshold signal at some time, it will retain that voltage boost forever until it fires again. This characteristic is clearly not in line with observed neuronal behavior.

Hodgkin–Huxley Model

Experimental evidence supporting the model	
Property of the H&H model	**References**
The shape of an individual spike	
The identity of the ions involved	
Spike speed across the axon	

The Hodgkin–Huxley model (H&H model) is a model of the relationship between ion currents crossing the neuronal cell membrane and the membrane voltage. The model is based on experiments that allowed to force membrane voltage using an intra-cellular pipette. This model is based on the concept of membrane ion channels and relies on data from the squid giant axon. In terms of recognition by the scientific community, this model is a very successful as Hodgkin–Huxley won the Nobel Prize for their work.

We note as before our voltage-current relationship, this time generalized to include multiple voltage-dependent currents:

$$C_m \frac{dV(t)}{dt} = -\sum_i I_i(t, V).$$

Each current is given by Ohm's Law as

$$I(t,V) = g(t,V) \cdot (V - V_{eq})$$

where $g(t,V)$ is the conductance, or inverse resistance, which can be expanded in terms of its constant average $\bar{g}$ and the activation and inactivation fractions m and h, respectively, that determine how many ions can flow through available membrane channels. This expansion is given by

$$g(t,V) = \bar{g} \cdot m(t,V)^p \cdot h(t,V)^q$$

and our fractions follow the first-order kinetics

$$\frac{dm(t,V)}{dt} = \frac{m_\infty(V) - m(t,V)}{\tau_m(V)} = \alpha_m(V) \cdot (1-m) - \beta_m(V) \cdot m$$

with similar dynamics for h, where we can use either τ and m_∞ or α and β to define our gate fractions.

With such a form, all that remains is to individually investigate each current one wants to include. Typically, these include inward Ca^{2+} and Na^+ input currents and several varieties of K^+ outward currents, including a "leak" current.

The end result can be at the small end 20 parameters which one must estimate or measure for an accurate model, and for complex systems of neurons not easily tractable by computer. Careful simplifications of the Hodgkin–Huxley model are therefore needed.

Leaky Integrate-and-fire

In the leaky integrate-and-fire model, the memory problem is solved by adding a "leak" term to the membrane potential, reflecting the diffusion of ions that occurs through the membrane when some equilibrium is not reached in the cell. The model looks like

$$I(t) - \frac{V_m(t)}{R_m} = C_m \frac{dV_m(t)}{dt}$$

where R_m is the membrane resistance, as we find it is not a perfect insulator as assumed previously. This forces the input current to exceed some threshold $I_{th} = V_{th} / R_m$ in order to cause the cell to fire, else it will simply leak out any change in potential. The firing frequency thus looks like

$$f(I) = \begin{cases} 0, & I \le I_{th} \\ [t_{ref} - R_m C_m \log(1 - \frac{V_{th}}{IR_m})]^{-1}, & I > I_{th} \end{cases}$$

which converges for large input currents to the previous leak-free model with refractory period.

Galves-Löcherbach

The Galves-Löcherbach model is a specific development of the leaky integrate-and-fire model. It is inherently stochastic. It was developed by mathematicians Antonio Galves and Eva Löcherbach. Given the model specifications, the probability that a given neuron i spikes in a time period t may be described by

$$\text{Prob}(X_t(i) = 1 \mid \mathcal{F}_{t-1}) = \phi_i\Big(\sum_{j\in I} W_{j\to i} \sum_{s=L_t^i}^{t-1} g_j(t-s)X_s(j), t - L_t^i\Big),$$

where $W_{j\to i}$ is a synaptic weight, describing the influence of neuron j on neuron i, g_j expresses the leak, and L_t^i provides the spiking history of neuron i before t, according to

$$L_t^i = \sup\{s < t : X_s(i) = 1\}.$$

Exponential Integrate-and-fire

In the Exponential Integrate-and-Fire, spike generation is exponential, following the equation:

$$\frac{dX}{dt} = \Delta_T \exp\left(\frac{X - X_T}{\Delta_T}\right).$$

where X is the membrane potential, X_T is the membrane potential threshold, and $\ddot{A}_T$ is the sharpness of action potential initiation, usually around 1 mV for cortical pyramidal neurons. Once the membrane potential crosses X_T, it diverges to infinity in finite time.

FitzHugh–Nagumo

Sweeping simplifications to Hodgkin–Huxley were introduced by FitzHugh and Nagumo in 1961 and 1962. Seeking to describe "regenerative self-excitation" by a nonlinear positive-feedback membrane voltage and recovery by a linear negative-feedback gate voltage, they developed the model described by

$$\frac{dV}{dt} = V - V^3 - w + I_{\text{ext}}$$

$$\tau\frac{dw}{dt} = V - a - bw$$

where we again have a membrane-like voltage and input current with a slower general gate voltage w and experimentally-determined parameters $a = -0.7$, $b = 0.8$, $\tau = 1/0.08$. Although not clearly derivable from biology, the model allows for a simplified, immediately available dynamic, without being a trivial simplification.

Morris–Lecar

In 1981 Morris and Lecar combined Hodgkin–Huxley and FitzHugh–Nagumo into a voltage-gated calcium channel model with a delayed-rectifier potassium channel, represented by

$$C \frac{dV}{dt} = -I_{ion}(V, w) + I$$

$$\frac{dw}{dt} = \phi \cdot \frac{w_\infty - w}{\tau_w}$$

where $I_{ion}(V, w) = \overline{g}_{Ca} m_\infty \cdot (V - V_{Ca}) + \overline{g}_K w \cdot (V - V_K) + \overline{g}_L \cdot (V - V_L)$.

Hindmarsh–Rose

Building upon the FitzHugh–Nagumo model, Hindmarsh and Rose proposed in 1984 a model of neuronal activity described by three coupled first order differential equations:

$$\frac{dx}{dt} = y + 3x^2 - x^3 - z + I$$

$$\frac{dy}{dt} = 1 - 5x^2 - y$$

$$\frac{dz}{dt} = r \cdot (4(x + \tfrac{8}{5}) - z)$$

with $r^2 = x^2 + y^2 + z^2$, and $r \approx 10^{-2}$ so that the z variable only changes very slowly. This extra mathematical complexity allows a great variety of dynamic behaviors for the membrane potential, described by the x variable of the model, which include chaotic dynamics. This makes the Hindmarsh–Rose neuron model very useful, because being still simple, allows a good qualitative description of the many different patterns of the action potential observed in experiments.

Cable Theory

Cable theory describes the dendritic arbor as a cylindrical structure undergoing a regular pattern of bifurcation, like branches in a tree. For a single cylinder or an entire tree, the input conductance at the base (where the tree meets the cell body, or any such boundary) is defined as

$$G_{in} = \frac{G_\infty \tanh(L) + G_L}{1 + (G_L / G_\infty)\tanh(L)},$$

where L is the electrotonic length of the cylinder which depends on its length, diameter, and resistance. A simple recursive algorithm scales linearly with the number of branches and can be used to calculate the effective conductance of the tree. This is given by

$$G_D = G_m A_D \tanh(L_D) / L_D$$

where $A_D = \pi l d$ is the total surface area of the tree of total length l, and L_D is its total electrotonic length. For an entire neuron in which the cell body conductance is G_S and the membrane conductance per unit area is $G_{md} = G_m / A$, we find the total neuron conductance G_N for n dendrite trees by adding up all tree and soma conductances, given by

$$G_N = G_S + \sum_{j=1}^{n} A_{D_j} F_{dga_j},$$

where we can find the general correction factor F_{dga} experimentally by noting $G_D = G_{m}$-$_d A_D F_{dga}$.

Compartmental Models

The cable model makes a number of simplifications to give closed analytic results, namely that the dendritic arbor must branch in diminishing pairs in a fixed pattern. A compartmental model allows for any desired tree topology with arbitrary branches and lengths, but makes simplifications in the interactions between branches to compensate. Thus, the two models give complementary results, neither of which is necessarily more accurate.

Each individual piece, or compartment, of a dendrite is modeled by a straight cylinder of arbitrary length l and diameter d which connects with fixed resistance to any number of branching cylinders. We define the conductance ratio of the ith cylinder as $B_i = G_i / G_\infty$, where $G_\infty = \dfrac{\pi d^{3/2}}{2\sqrt{R_i R_m}}$ and R_i is the resistance between the current compartment and the next. We obtain a series of equations for conductance ratios in and out of a compartment by making corrections to the normal dynamic $B_{out,i} = B_{in,i+1}$, as

$$B_{out,i} = \frac{B_{in,i+1}(d_{i+1} / d_i)^{3/2}}{\sqrt{R_{m,i+1} / R_{m,i}}}$$

$$B_{in,i} = \frac{B_{out,i} + \tanh X_i}{1 + B_{out,i} \tanh X_i}$$

$$B_{out,par} = \frac{B_{in,dau1}(d_{dau1}/d_{par})^{3/2}}{\sqrt{R_{m,dau1}/R_{m,par}}} + \frac{B_{in,dau2}(d_{dau2}/d_{par})^{3/2}}{\sqrt{R_{m,dau2}/R_{m,par}}} + \dots$$

where the last equation deals with *parents* and *daughters* at branches, and $X_i = \dfrac{l_i\sqrt{4R_i}}{\sqrt{d_iR_m}}$.

We can iterate these equations through the tree until we get the point where the dendrites connect to the cell body (soma), where the conductance ratio is $B_{in,stem}$. Then our total neuron conductance is given by

$$G_N = \frac{A_{soma}}{R_{m,soma}} + \sum_j B_{in,stem,j} G_{\infty,j}.$$

An example of a compartmental model of a neuron, with an algorithm to reduce the number of compartments (increase the computational speed) and yet retain the salient electrical characteristics, can be found in.

Natural Input Stimulus Neuron Models

The models in this category were derived following experiments involving natural stimulation such as light, sound, touch, or odor. In these experiments, the spike pattern resulting from each stimulus presentation varies from trial to trial, but the averaged response from several trials often converges to a clear pattern. Consequently, the models in this category generate a probabilistic relationship between the input stimulus to spike occurrences.

The Non-homogeneous Poisson Process Model (Siebert)

Siebert modeled the neuron spike firing pattern using a non-homogeneous Poisson process model, following experiments involving the auditory system. According to Siebert, the probability of a spiking event at the time interval $[t, t + \Delta_t]$ is proportional to a non negative function $g[s(t)]$, where $s(t)$ is the raw stimulus.:

$$P_{spike}(t \in [t', t' + \Delta_t]) = \Delta_t \cdot g[s(t)]$$

Siebert considered several functions as $g[s(t)]$, including $g[s(t)] \propto s^2(t)$ for low stimulus intensities.

The main advantage of Siebert's model is its simplicity. The shortcomings of the model is its inability to reflect properly the following phenomena:

- The edge emphasizing property of the neuron in response to a stimulus pulse.

- The saturation of the firing rate.

- The values of inter-spike-interval-histogram at short intervals values (close to zero).

These shortcoming are addressed by the two state Markov Model.

The Two State Markov Model (Nossenson & Messer)

The spiking neuron model by Nossenson & Messer produces the probability of the neuron to fire a spike as a function of either an external or pharmacological stimulus. The model consists of a cascade of a receptor layer model and a spiking neuron model, as shown in Fig 4. The connection between the external stimulus to the spiking probability is made in two steps: First, a receptor cell model translates the raw external stimulus to neurotransmitter concentration, then, a spiking neuron model connects between neurotransmitter concentration to the firing rate (spiking probability). Thus, the spiking neuron model by itself depends on neurotransmitter concentration at the input stage.

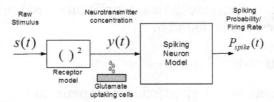

Fig 4: High level block diagram of the receptor layer and neuron model by Nossenson & Messer.

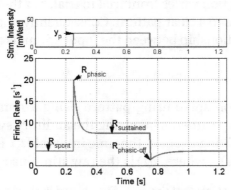

Fig 5. The prediction for the firing rate in response to a pulse stimulus as given by the model by Nossenson & Messer.

An important feature of this model is the prediction for neurons firing rate pattern which captures, using a low number of free parameters, the characteristic edge emphasized response of neurons to a stimulus pulse, as shown in Fig. 5. The firing rate is identified both as a normalized probability for neural spike firing, and as a quantity proportional to the current of neurotransmitters released by the cell. The expression for the firing rate takes the following form:

$$R_{fire}(t) = \frac{P_{spike}(t;\Delta_t)}{\Delta_t} = [y(t) + R_0] \cdot P_0(t)$$

where,

- P_0 is the probability of the neuron to be "armed" and ready to fire. It is given by the following differential equation:

$$\dot{P_0} = -[y(t) + R_0 + R_1] \cdot P_0(t) + R_1$$

P_0 could be generally calculated recursively using Euler method, but in the case of a pulse of stimulus it yields a simple closed form expression.

- $y(t)$ is the input of the model and is interpreted as the neurotransmitter concentration on the cell surrounding (in most cases glutamate) . For an external stimulus it can be estimated through the receptor layer model:

$y(t) \simeq g_{gain} \cdot \langle s^2(t) \rangle$, with $\langle s^2(t) \rangle$ being short temporal average of stimulus power (given in Watt or other energy per time unit).

- R_0 corresponds to the intrinsic spontaneous firing rate of the neuron.

- R_1 is the recovery rate of the neuron from refractory state.

Other predictions by this model include:

1) The averaged Evoked Response Potential (ERP) due to population of many neurons in unfiltered measurements resembles the firing rate.

2) The voltage variance of activity due to multiple neuron activity resembles the firing rate (also known as Multi-Unit-Activity power or MUA).

3) The inter-spike-interval probability distribution takes the form a gamma-distribution like function.

Experimental evidence supporting the model by Nossenson & Messer		
Property of the Model by Nossenson & Messer	**References**	**Description of experimental evidence**
The shape of the firing rate in response to an auditory stimulus pulse		The Firing Rate has the same shape of Fig 5.
The shape of the firing rate in response to a visual stimulus pulse		The Firing Rate has the same shape of Fig 5.
The shape of the firing rate in response to an olfactory stimulus pulse		The Firing Rate has the same shape of Fig 5.

Experimental evidence supporting the model by Nossenson & Messer		
Property of the Model by Nossenson & Messer	**References**	**Description of experimental evidence**
The shape of the firing rate in response to a somato-sensory stimulus		The Firing Rate has the same shape of Fig 5.
The change in firing rate in response to neurotransmitter application (mostly glutamate)		Firing Rate change in response to neurotransmitter application (Glutamate)
Square dependence between an auditory stimulus pressure and the firing rate		Square Dependence between Auditory Stimulus pressure and the Firing Rate (- Linear dependence in pressure square (power)).
Square dependence between visual stimulus electric field (volts) and the firing rate		Square dependence between visual stimulus electric field (volts) - Linear Dependence between Visual Stimulus *Power* and the Firing Rate.
The shape of the Inter-Spike-Interval Statistics (ISI)		ISI shape resembles the gamma-function-like
The ERP resembles the firing rate in unfiltered measurements		The shape of the averaged evoked response potential in response to stimulus resembles the firing rate (Fig. 5).
MUA power resembles the firing rate		The shape of the empirical variance of extra-cellular measurements in response to stimulus pulse resembles the firing rate (Fig. 5).

Non-Markovian Models

The following is a list of published non-Markovian neuron models:

- Johnson, and Swami

- Berry and Meister

- Kass and Ventura

Pharmacological Input Stimulus Neuron Models

The models in this category produce predictions for experiments involving pharmacological stimulation.

Synaptic Transmission (Koch & Segev)

According to the model by Koch and Segev, the response of a neuron to individual neurotransmitters can be modeled as an extension of the classical Hodgkin–Huxley model with both standard and nonstandard kinetic currents. Four neurotransmitters primarily have influence in the CNS. AMPA/kainate receptors are fast excitatory mediators while NMDA receptors mediate considerably slower currents. Fast inhibitory currents go through $GABA_A$ receptors, while $GABA_B$ receptors mediate by secondary G-protein-activated potassium channels. This range of mediation produces the following current dynamics:

- $I_{AMPA}(t,V) = \bar{g}_{AMPA} \cdot [O] \cdot (V(t) - E_{AMPA})$

- $I_{NMDA}(t,V) = \bar{g}_{NMDA} \cdot B(V) \cdot [O] \cdot (V(t) - E_{NMDA})$

- $I_{GABA_A}(t,V) = \bar{g}_{GABA_A} \cdot ([O_1] + [O_2]) \cdot (V(t) - E_{Cl})$

- $I_{GABA_B}(t,V) = \bar{g}_{GABA_B} \cdot \frac{[G]^n}{[G]^n + K_d} \cdot (V(t) - E_K)$

where $\bar{g}$ is the maximal conductance (around 1S) and E is the equilibrium potential of the given ion or transmitter (AMDA, NMDA, Cl, or K), while $[O]$ describes the fraction of receptors that are open. For NMDA, there is a significant effect of *magnesium block* that depends sigmoidally on the concentration of intracellular magnesium by $B(V)$. For $GABA_B$, $[G]$ is the concentration of the G-protein, and K_d describes the dissociation of G in binding to the potassium gates.

The dynamics of this more complicated model have been well-studied experimentally and produce important results in terms of very quick synaptic potentiation and depression, that is, fast, short-term learning.

The Two State Markov Model (Nossenson & Messer)

The model by Nossenson and Messer translates neurotransmitter concentration at the input stage to the probability of releasing neurotransmitter at the output stage. For a more detailed description of this model, see the Two state Markov model section above.

Conjectures Regarding the Role of the Neuron in the Wider Context of the Brain Principle of Operation

Conjecture 1: Relation between Artificial and Biological Neuron Models

The most basic model of a neuron consists of an input with some synaptic weight vector and an activation function or transfer function inside the neuron determining output. This is the basic structure used in artificial neurons, which in a neural network often looks like

$$y_i = \phi\left(\sum_j w_{ij} x_j\right)$$

where y_i is the output of the i th neuron, x_j is the jth input neuron signal, w_{ij} is the synaptic weight (or strength of connection) between the neurons i and j, and φ is the activation function. While this model has seen success in machine-learning applications, it is a poor model for real (biological) neurons, because it lacks the time-dependence that real neurons exhibit. Some of the earliest biological models took this form until kinetic models such as the Hodgkin–Huxley model became dominant.

In the case of modelling a biological neuron, physical analogues are used in place of abstractions such as "weight" and "transfer function". A neuron is filled and surrounded with water containing ions, which carry electric charge. The neuron is bound by an insulating cell membrane and can maintain a concentration of charged ions on either side that determines a capacitance C_m. The firing of a neuron involves the movement of ions into the cell that occurs when neurotransmitters cause ion channels on the cell membrane to open. We describe this by a physical time-dependent current $I(t)$. With this comes a change in voltage, or the electrical potential energy difference between the cell and its surroundings, which is observed to sometimes result in a voltage spike called an action potential which travels the length of the cell and triggers the release of further neurotransmitters. The voltage, then, is the quantity of interest and is given by $V_m(t)$.

Conjecture 2: Loops of Spiking Neurons for Decision Making

Conjecture 3: The Neurotransmitter Based Energy Detection Scheme

The neurotransmitter based energy detection scheme suggests that the neural tissue chemically executes a Radar-like detection procedure. A list of experimental evidence supporting this conjecture is given in. This conjecture attributes active functional roles to non-spiking neurons and glia cells.

General Comments Regarding the Modern Perspective of Scientific and Engineering Models

- The models above are still idealizations. Corrections must be made for the increased membrane surface area given by numerous dendritic spines, temperatures significantly hotter than room-temperature experimental data, and non-uniformity in the cell's internal structure. Certain observed effects do not fit into some of these models. For instance, the temperature cycling (with minimal net temperature increase) of the cell membrane during action potential propagation not compatible with models which rely on modeling the membrane as a resistance which must dissipate energy when current flows through it. The transient thickening of the cell membrane during action potential propagation is also not predicted by these models, nor is the changing capacitance and voltage spike that results from this thickening incorporated into these models. The action of some anesthetics such as inert gases is problematic for these models as well. New models, such as the soliton model attempt to explain these phenomena, but are less developed than older models and have yet to be widely applied. Also improbable possibility of modelling of local chronobiology mechanisms.

- Modern views regarding of the role of the scientific model suggest that "All models are wrong but some are useful" (Box and Draper, 1987, Gribbin, 2009; Paninski et al., 2009).

Hodgkin–Huxley Model

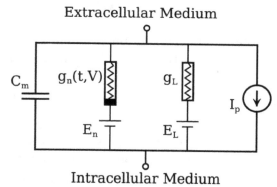

Extracellular Medium

Intracellular Medium

Basic components of Hodgkin–Huxley-type models. Hodgkin–Huxley type models represent the biophysical characteristic of cell membranes. The lipid bilayer is represented as a capacitance (C_m). Voltage-gated and leak ion channels are represented by nonlinear (g_n) and linear (g_L) conductances, respectively. The electrochemical gradients driving the flow of ions are represented by batteries (E), and ion pumps and exchangers are represented by current sources (I_p).

The Hodgkin–Huxley model, or conductance-based model, is a mathematical model that describes how action potentials in neurons are initiated and propagated. It is a set of nonlinear differential equations that approximates the electrical characteristics of excitable cells such as neurons and cardiac myocytes, and hence it is a continuous time model, unlike the Rulkov map for example.

Alan Lloyd Hodgkin and Andrew Fielding Huxley described the model in 1952 to explain the ionic mechanisms underlying the initiation and propagation of action potentials in the squid giant axon. They received the 1963 Nobel Prize in Physiology or Medicine for this work.

Basic Components

The typical Hodgkin–Huxley model treats each component of an excitable cell as an electrical element. The lipid bilayer is represented as a capacitance (C_m). Voltage-gated ion channels are represented by electrical conductances (g_n, where n is the specific ion channel) that depend on both voltage and time. Leak channels are represented by linear conductances (g_L). The electrochemical gradients driving the flow of ions are represented by voltage sources (E_n) whose voltages are determined by the ratio of the intra- and extracellular concentrations of the ionic species of interest. Finally, ion pumps are represented by current sources (I_p). The membrane potential is denoted by V_m.

Mathematically, the current flowing through the lipid bilayer is written as

$$I_c = C_m \frac{dV_m}{dt}$$

and the current through a given ion channel is the product

$$I_i = g_n (V_m - V_i)$$

where V_i is the reversal potential of the i-th ion channel. Thus, for a cell with sodium and potassium channels, the total current through the membrane is given by:

$$I = C_m \frac{dV_m}{dt} + g_K (V_m - V_K) + g_{Na} (V_m - V_{Na}) + g_l (V_m - V_l)$$

where I is the total membrane current per unit area, C_m is the membrane capacitance per unit area, g_K and g_{Na} are the potassium and sodium conductances per unit area, respectively, V_K and V_{Na} are the potassium and sodium reversal potentials, respectively, and g_l and V_l are the leak conductance per unit area and leak reversal potential, respectively. The time dependent elements of this equation are V_m, g_{Na}, and g_K, where the last two conductances depend explicitly on voltage as well.

Ionic Current Characterization

In voltage-gated ion channels, the channel conductance g_i is a function of both time and voltage ($g_n(t, V)$ in the figure), while in leak channels g_i is a constant (g_L in the figure). The current generated by ion pumps is dependent on the ionic species specific to that pump. The following sections will describe these formulations in more detail.

Voltage-gated Ion Channels

Using a series of voltage clamp experiments and by varying extracellular sodium and potassium concentrations, Hodgkin and Huxley developed a model in which the properties of an excitable cell are described by a set of four ordinary differential equations. Together with the equation for the total current mentioned above, these are:

$$I = C_m \frac{dV_m}{dt} + \bar{g}_K n^4 (V_m - V_K) + \bar{g}_{Na} m^3 h (V_m - V_{Na}) + \bar{g}_l (V_m - V_l),$$

$$\frac{dn}{dt} = \alpha_n(V_m)(1-n) - \beta_n(V_m)n$$

$$\frac{dm}{dt} = \alpha_m(V_m)(1-m) - \beta_m(V_m)m$$

$$\frac{dh}{dt} = \alpha_h(V_m)(1-h) - \beta_h(V_m)h$$

where I is the current per unit area, and α_i and β_i are rate constants for the i-th ion channel, which depend on voltage but not time. $\bar{g}_n$ is the maximal value of the conductance. n, m, and h are dimensionless quantities between 0 and 1 that are associated with potassium channel activation, sodium channel activation, and sodium channel inactivation, respectively. For $p = (n,m,h)$, α_p and β_p take the form

$$\alpha_p(V_m) = p_\infty(V_m) / \tau_p$$

$$\beta_p(V_m) = (1 - p_\infty(V_m)) / \tau_p.$$

p_∞ and $(1-p_\infty)$ are the steady state values for activation and inactivation, respectively, and are usually represented by Boltzmann equations as functions of V_m. In the original paper by Hodgkin and Huxley, the functions α and β are given by

$$\alpha_n(V_m) = \frac{0.01(V_m+10)}{\exp(\frac{V_m+10}{10})-1} \quad \alpha_m(V_m) = \frac{0.1(V_m+25)}{\exp(\frac{V_m+25}{10})-1} \quad \alpha_h(V_m) = 0.07\exp(\frac{V_m}{20})$$

$$\beta_n(V_m) = 0.125\exp(\frac{V_m}{80}) \quad \beta_m(V_m) = 4\exp(\frac{V_m}{18}) \quad \beta_h(V_m) = \frac{1}{\exp(\frac{V_m+30}{10})+1}$$

while in many current software programs, Hodgkin–Huxley type models generalize α and β to

$$\frac{A_p(V_m - B_p)}{\exp(\frac{V_m - B_p}{C_p}) - D_p}$$

In order to characterize voltage-gated channels, the equations are fit to voltage clamp data. For a derivation of the Hodgkin–Huxley equations under voltage-clamp, see. Briefly, when the membrane potential is held at a constant value (i.e., voltage-clamp), for each value of the membrane potential the nonlinear gating equations reduce to equations of the form:

$$m(t) = m_0 - [(m_0 - m_\infty)(1 - e^{-t/\tau_m})]$$

$$h(t) = h_0 - [(h_0 - h_\infty)(1 - e^{-t/\tau_h})]$$

$$n(t) = n_0 - [(n_0 - n_\infty)(1 - e^{-t/\tau_n})]$$

Thus, for every value of membrane potential V_m the sodium and potassium currents can be described by

$$I_{Na}(t) = \bar{g}_{Na} m(V_m)^3 h(V_m)(V_m - E_{Na}),$$

$$I_K(t) = \bar{g}_K n(V_m)^4 (V_m - E_K).$$

In order to arrive at the complete solution for a propagated action potential, one must write the current term I on the left-hand side of the first differential equation in terms of V, so that the equation becomes an equation for voltage alone. The relation between I and V can be derived from cable theory and is given by

$$I = \frac{a}{2R} \frac{\partial^2 V}{\partial x^2},$$

where a is the radius of the axon, R is the specific resistance of the axoplasm, and x is the position along the nerve fiber. Substitution of this expression for I transforms the original set of equations into a set of partial differential equations, because the voltage becomes a function of both x and t.

The Levenberg–Marquardt algorithm, a modified Gauss–Newton algorithm, is often used to fit these equations to voltage-clamp data.

While the original experiments treated only sodium and potassium channels, the Hodgkin Huxley model can also be extended to account for other species of ion channels.

Leak Channels

Leak channels account for the natural permeability of the membrane to ions and take the form of the equation for voltage-gated channels, where the conductance g_i is a constant.

Pumps and Exchangers

The membrane potential depends upon the maintenance of ionic concentration gradients across it. The maintenance of these concentration gradients requires active transport of ionic species. The sodium-potassium and sodium-calcium exchangers are the best known of these. Some of the basic properties of the Na/Ca exchanger have already been well-established: the stoichiometry of exchange is 3 Na^+: 1 Ca^{2+} and the exchanger is electrogenic and voltage-sensitive. The Na/K exchanger has also been described in detail, with a 3 Na^+: 2 K^+ stoichiometry.

Mathematical Properties

The Hodgkin–Huxley model can be thought of as a differential equation with four state variables, $v(t)$, $m(t)$, $n(t)$, and $h(t)$, that change with respect to time t. The system is difficult to study because it is a nonlinear system and cannot be solved analytically. However, there are many numeric methods available to analyze the system. Certain properties and general behaviors, such as limit cycles, can be proven to exist.

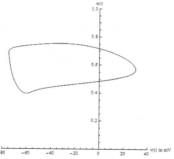

A simulation of the Hodgkin–Huxley model in phase space, in terms of voltage v(t) and potassium gating variable n(t). The closed curve is known as a limit cycle.

Center Manifold

Because there are four state variables, visualizing the path in phase space can be difficult. Usually two variables are chosen, voltage $v(t)$ and the potassium gating variable $n(t)$, allowing one to visualize the limit cycle. However, one must be careful because this is an ad-hoc method of visualizing the 4-dimensional system. This does not prove the existence of the limit cycle.

A better projection can be constructed from a careful analysis of the Jacobian of the system, evaluated at the equilibrium point. Specifically, the eigenvalues of the Jacobian are indicative of the center manifold's existence. Likewise, the eigenvectors of the Jacobian reveal the center manifold's orientation. The Hodgkin–Huxley model has two negative eigenvalues and two complex eigenvalues with slightly positive real parts. The eigenvectors associated with the two negative eigenvalues will reduce to zero as time t increases. The remaining two complex eigenvectors define the center manifold. In other words, the 4-dimensional system collapses onto a 2-dimensional plane. Any solution starting off the center manifold will decay towards the center manifold. Furthermore, the limit cycle is contained on the center manifold.

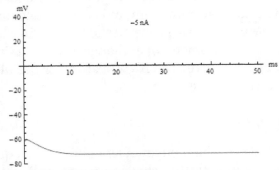

The voltage $v(t)$ (in millivolts) of the Hodgkin–Huxley model, graphed over 50 milliseconds. The injected current varies from −5 nanoamps to 12 nanoamps. The graph passes through three stages: an equilibrium stage, a single-spike stage, and a limit cycle stage.

Bifurcations

If we use the injected current as a bifurcation parameter, then the Hodgkin–Huxley model undergoes a Hopf bifurcation. As with most neuronal models, increasing the injected current will increase the firing rate of the neuron. One consequence of the Hopf bifurcation is that there is a minimum firing rate. This means that either the neuron is not firing at all (corresponding to zero frequency), or firing at the minimum firing rate. Because of the all or none principle, there is no smooth increase in action potential amplitude, but rather there is a sudden "jump" in amplitude. The resulting transition is known as a classical canard phenomenon, or simply a canard.

Improvements and Alternative Models

The Hodgkin–Huxley model is regarded as one of the great achievements of 20th-century biophysics. Nevertheless, modern Hodgkin–Huxley-type models have been extended in several important ways:

- Additional ion channel populations have been incorporated based on experimental data.

- The Hodgkin–Huxley model has been modified to incorporate transition state theory and produce thermodynamic Hodgkin–Huxley models.

- Models often incorporate highly complex geometries of dendrites and axons, often based on microscopy data.

- Stochastic models of ion-channel behavior, leading to stochastic hybrid systems

Several simplified neuronal models have also been developed (such as the FitzHugh–

Nagumo model), facilitating efficient large-scale simulation of groups of neurons, as well as mathematical insight into dynamics of action potential generation.

Binding Neuron

A binding neuron (BN) is an abstract mathematical model of the electrical activity of a neuron, closely related to well-known integrate-and-fire model. The BN model originated in a 1998 paper by A. K. Vidybida

Description of the Concept

For a generic neuron the stimuli are excitatory impulses. Normally, more than single input impulse is necessary for exciting neuron up to the level when it fires and emits an output impulse. Let the neuron receives n input impulses at consecutive moments of time $t_1, t_2, \ldots, t_n$. In the BN concept the temporal coherence tc between input impulses is defined as follows

$$tc = \frac{1}{t_n - t_1}.$$

The high degree of temporal coherence between input impulses suggests that in external media all n impulses can be created by a single complex event. Correspondingly, if BN is stimulated by a highly coherent set of input impulses, it fires and emits an output impulse. In the BN terminology we say that BN binds the elementary events (input impulses) into a single event (output impulse). The binding happens if the input impulses are enough coherent in time, and does not happen if those impulses do not have required degree of coherence.

Inhibition in the BN concept (essentially, the slow somatic potassium inhibition) controls the degree of temporal coherence required for binding: the higher level of inhibition, the higher degree of temporal coherence is necessary for binding to occur.

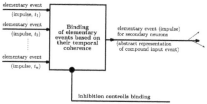

Scheme of signal processing in accordance with binding neuron concept. $t_1, t_2, \ldots, t_n$ --- the moments of receiving of input impulses.

The emitted output impulse is treated as abstract representation of the compound event (the set of coherent in time input impulses).

Origin

"Although a neuron requires energy, its main function is to receive signals and to send them out that is, to handle information." --- this words by Francis Crick point at the necessity to describe neuronal functioning in terms of processing of abstract signals The two abstract concepts, namely, the "coincidence detector" and "temporal integrator" are offered in this course, . The first one expects that a neuron fires a spike if a number of input impulses are received at the same time. In the temporal integrator concept a neuron fires a spike after receiving a number of input impulses distributed in time. Each of the two takes into account some features of real neurons since it is known that a realistic neuron can display both coincidence detector and temporal integrator modes of activity depending on the stimulation applied, . At the same time, it is known that a neuron together with excitatory impulses receives also inhibitory stimulation. A natural development of the two above mentioned concepts could be a concept which endows inhibition with its own signal processing role.

In the neuroscience, there is an idea of binding problem. For example, during visual perception, such features as form, color and stereopsis are represented in the brain by different neuronal assemblies. The mechanism ensuring those features to be perceived as belonging to a single real object is called "feature binding", . The experimentally approved opinion is that precise temporal coordination between neuronal impulses is required for the binding to occur, , , , , . This coordination mainly means that signals about different features must arrive to certain areas in the brain within a certain time window.

The BN concept reproduces at the level of single generic neuron the requirement, which is necessary for the feature binding to occur, and which was formulated earlier at the level of large-scale neuronal assemblies. Its formulation is made possible by the analysis of response of the Hodgkin–Huxley model to stimuli similar to those the real neurons receive in the natural conditions.

Integrated Circuit Implementation

The above-mentioned and other neuronal models and nets of them can be implement-ed in microchips. Among different chips it is worth mentioning the FPGA ones. The FPGA chips can be used for implementation of any neuronal model, but the BN model can be programmed most naturally because it can use only integers and do not need solving differential equations. Those features are used, e.g. in .

Limitations

As an abstract concept the BN one is subjected to necessary limitations. Among those are such as ignoring neuronal morphology, identical magnitude of input impulses, re-placement of a set of transients with different relaxation times, known for a real neu-ron, with a single time to live, τ, of impulse in neuron, the absence of refractoriness and fast (chlorine) inhibition. The same limitations has the BN model, yet some of them can be removed in a complicated model, e.g. , where the BN model is used with refractorines and fast inhibition.

Efficient Coding Hypothesis

The efficient coding hypothesis was proposed by Horace Barlow in 1961 as a theoretical model of sensory coding in the brain. Within the brain, neurons often communicate with one another by sending electrical impulses referred to as action potentials or spikes. One goal of sensory neuroscience is to decipher the meaning of these spikes in order to understand how the brain represents and processes information about the outside world. Barlow hypothesized that the spikes in the sensory system formed a neural code for efficiently representing sensory information. By efficient Barlow meant that the code minimized the number of spikes needed to transmit a given signal. This is somewhat analogous to transmitting information across the internet, where different file formats can be used to transmit a given image. Different file formats require different number of bits for representing the same image at given distortion level, and some are better suited for representing certain classes of images than others. According to this model, the brain is thought to use a code which is suited for representing visual and audio information representative of an organism's natural environment.

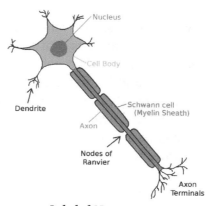

Labeled Neuron

Efficient Coding and Information Theory

The development of the Barlow's hypothesis was influenced by information theory introduced by Claude Shannon only a decade before. Information theory provides the mathematical framework for analyzing communication systems. It formally defines concepts such as information, channel capacity, and redundancy. Barlow's model treats the sensory pathway as a communication channel where neuronal spiking is an efficient code for representing sensory signals. The spiking code aims to maximize available channel capacity by minimizing the redundancy between representational units.

A key prediction of the efficient coding hypothesis is that sensory processing in the brain should be adapted to natural stimuli. Neurons in the visual (or auditory) system should be optimized for coding images (or sounds) representative of those found in nature. Researchers have shown that filters optimized for coding natural images lead to filters which resemble the receptive fields of simple-cells in V1. In the auditory domain, optimizing a network for coding natural sounds leads to filters which resemble the impulse response of cochlear filters found in the inner ear.

Constraints on the Visual System

Due to constraints on the visual system such as the number of neurons and the metabolic energy required for "neural activities", the visual processing system must have an efficient strategy for transmitting as much information as possible. Information must be compressed as it travels from the retina back to the visual cortex. While the retinal receptors can receive information at 10^9 bit/s, the optic nerve, which is composed of 1 million ganglion cells transmitting at 1 bit/sec, only has a transmission capacity of 10^6 bit/s. Further reduction occurs that limits the overall transmission to 40 bit/s which results in inattentional blindness. Thus, the hypothesis states that neurons should encode information as efficiently as possible in order to maximize neural resources. For example, it has been shown that visual data can be compressed up to 20 fold without noticeable information loss.

Evidence suggests that our visual processing system engages in bottom-up selection. For example, inattentional blindness suggests that there must be data deletion early on in the visual pathway. This bottom-up approach allows us to respond to unexpected and salient events more quickly and is often directed by attentional selection. This also gives our visual system the property of being goal-directed. Many have suggested that the visual system is able to work efficiently by breaking images down into distinct components. Additionally, it has been argued that the visual system takes advantage of redundancies in inputs in order to transmit as much information as possible while using the fewest resources.

Evolution-based Neural System

Simoncelli and Olshausen outline the three major concepts that are assumed to be involved in the development of systems neuroscience:

1. an organism has specific tasks to perform

2. neurons have capabilities and limitations

3. an organism is in a particular environment.

One assumption used in testing the Efficient Coding Hypothesis is that neurons must be evolutionarily and developmentally adapted to the natural signals in their environment. The idea is that perceptual systems will be the quickest when responding to "environmental stimuli". The visual system should cut out any redundancies in the sensory input.

Natural Images and Statistics

Central to Barlow's hypothesis is information theory, which when applied to neuroscience, argues that an efficiently coding neural system "should match the statistics of the signals they represent". Therefore, it is important to be able to determine the statistics of the natural images that are producing these signals. Researchers have looked at various components of natural images including luminance contrast, color, and how images are registered over time. They can analyze the properties of natural scenes via digital cameras, spectrophotometers, and range finders.

Researchers look at how luminance contrasts are spatially distributed in an image: the luminance contrasts are highly correlated the closer they are in measurable distance and less correlated the farther apart the pixels are. independent component analysis (ICA) is an algorithm system that attempts to "linearly transform given (sensory) inputs into independent outputs (synaptic currents) ". ICA eliminates the redundancy by decorrelating the pixels in a natural image. Thus the individual components that make up the natural image are rendered statistically independent. However, researchers have thought that ICA is limited because it assumes that the neural response is linear, and therefore insufficiently describes the complexity of natural images. They argue that, despite what is assumed under ICA, the components of the natural image have a "higher-order structure" that involves correlations among components. Instead, researches have now developed temporal independent component analysis (TICA), which better represents the complex correlations that occur between components in a natural image. Additionally, a "hierarchical covariance model" developed by Karklin and Lewicki expands on sparse coding methods and can represent additional components of natural images such as "object location, scale, and texture".

The chromatic spectra as it comes from natural light, but also as it is reflected off of "natural materials" can be easily characterized with principal components analysis (PCA). Because the cones are absorbing a specific amount of photons from the natural image, researchers can use cone responses as a way of describing the natural image. Researchers have found that the three classes of cone receptors in the retina can accurately code natural images and that color is decorrelated already in the LGN. Time has also been modeled: natural images transform over time, and we can use these transformations to see how the visual input changes over time.

Hypotheses for Testing the Efficient Coding Hypothesis

If neurons are encoding according to the efficient coding hypothesis then individual neurons must be expressing their full output capacity. Before testing this hypothesis it is necessary to define what is considered to be a neural response. Simoncelli and Olshausen suggest that an efficient neuron needs to be given a maximal response value so that we can measure if a neuron is efficiently meeting the maximum level. Secondly, a population of neurons must not be redundant in transmitting signals and must be statistically independent. If the efficient coding hypothesis is accurate, researchers should observe is that there is sparsity in the neuron responses: that is, only a few neurons at a time should fire for an input.

Methodological Approaches for Testing the Hypotheses

One approach is to design a model for early sensory processing based on the statistics of a natural image and then compare this predicted model to how real neurons actually respond to the natural image. The second approach is to measure a neural system responding to a natural environment, and analyze the results to see if there are any statistical properties to this response.

Examples of these Approaches

1. Predicted Model Approach

In one study by Doi et al. in 2012, the researchers created a predicted response model of the retinal ganglion cells that would be based on the statistics of the natural images used, while considering noise and biological constraints. They then compared the actual information transmission as observed in real retinal ganglion cells to this optimal model to determine the efficiency. They found that the information transmission in the retinal ganglion cells had an overall efficiency of about 80% and concluded that "the functional connectivity between cones and retinal ganglion cells exhibits unique spatial structure...consistent with coding efficiency.

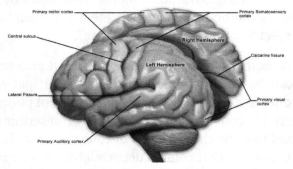

Primary visual cortex location on the right side

A study by van Hateren and Ruderman in 1998 used ICA to analyze video-sequnces and compared how a computer analyzed the independent components of the image to data for visual processing obtained from a cat in DeAngelis et al. 1993. The researchers described the independent components obtained from a video sequence as the "basic building blocks of a signal", with the independent component filter (ICF) measuring "how strongly each building block is present". They hypothesized that if simple cells are organized to pick out the "underlying structure" of images over time then cells should act like the independent component filters. They found that the ICFs determined by the computer were similar to the "receptive fields" that were observed in actual neurons.

2. Analyzing Actual Neural System in Response to Natural Images

In a report in *Science* from 2000, William E. Vinje and Jack Gallant outlined a series of experiments used to test elements of the efficient coding hypothesis, including a theory that the nCRF decorrelates projections from the primary visual cortex. To test this, they took recordings from the V1 neurons in awake macaques during "free viewing of natural images and conditions" that simulated natural vision conditions. The researchers hypothesized that the V1 uses sparse code, which is minimally redundant and "metabolically more efficient". They also hypothesized that interactions between the classical receptive field (CRF) and the non-classical receptive field (nCRF) produced this pattern of sparse coding during the viewing of these natural scenes. In order to test this, they created eye-scan paths and also extracted patches that ranged in size from 1-4 times the diameter of the CRF. They found that the sparseness of the coding increased with the size of the patch. Larger patches encompassed more of the nCRF—indicating that the interactions between these two regions created sparse code. Additionally as stimulus size increased, so did the sparseness. This suggests that the V1 uses sparse code when natural images span the entire visual field. The CRF was defined as the circular area surrounding the locations where stimuli evoked action potentials. They also tested to see if the stimulation of the nCRF increased the independence of the responses from the V1 neurons by randomly selecting pairs of neurons. They found that indeed, the neurons were more greatly decoupled upon stimulation of the nCRF. In conclusion, the experiments of Vinje and Gallant showed that the V1 uses sparse code by employing both the CRF and nCRF when viewing natural images, with the nCRF showing a definitive decorrelating effect on neurons which may increase their efficiency by increasing the amount of independent information they carry. They propose that the cells may represent the individual components of a given natural scene, which may contribute to pattern recognition

Another study done by Baddeley et al. had shown that firing-rate distributions of cat visual area V1 neurons and monkey inferotemporal (IT) neurons were exponential under naturalistic conditions, which implies optimal information transmission for a fixed average rate of firing. A subsequent study of monkey IT neurons found that only a minority were well described by an exponential firing distribution. De Polavieja later argued that this discrepancy was due to the fact that the exponential solution is correct only for the noise-free case, and showed that by taking noise into consideration, one could account for the observed results.

A study by Dan, Attick, and Reid in 1996 used natural images to test the hypothesis that early on in the visual pathway, incoming visual signals will be decorrelated to optimize efficiency. This decorrelation can be observed as the '"whitening" of the temporal and spatial power spectra of the neuronal signals". The researchers played natural image movies in front of cats and used a multielectrode array to record neural signals. This was achieved by refracting the eyes of the cats and then contact lenses being fitted into them. They found that in the LGN, the natural images were decorrelated and concluded, "the early visual pathway has specifically adapted for efficient coding of natural visual information during evolution and/or development".

Extensions

One of the implications of the efficient coding hypothesis is that the neural coding depends upon the statistics of the sensory signals. These statistics are a function of not only the environment (e.g., the statistics of the natural environment), but also the organism's behavior (e.g., how it moves within that environment). However, perception and behavior are closely intertwined in the perception-action cycle. For example, the process of vision involves various kinds of eye movements. An extension to the efficient coding hypothesis called active efficient coding (AEC) extends efficient coding to active perception. It hypothesizes that biological agents optimize not only their neural coding, but also their behavior to contribute to an efficient sensory representation of the environment. Along these lines, models for the development of active binocular vision and active visual tracking have been proposed.

Criticisms

Researchers should consider how the visual information is used: The hypothesis does not explain how the information from a visual scene is used—which is the main purpose of the visual system. It seems necessary to understand why we are processing image statistics from the environment because this may be relevant to how this information is ultimately processed. However, some researchers may see the irrelevance of the purpose of vision in Barlow's theory as an advantage for designing experiments.

Some experiments show correlations between neurons: When considering multiple neurons at a time, recordings "show correlation, synchronization, or other forms of statistical dependency between neurons". However, it is relevant to note that most of these experiments did not use natural stimuli to provoke these responses: this may not fit in directly to the efficient coding hypothesis because this hypothesis is concerned with natural image statistics. In his review article Simoncelli notes that perhaps we can interpret redundancy in the Efficient Coding Hypothesis a bit differently: he argues that statistical dependency could be reduced over "successive stages of processing", and not just in one area of the sensory pathway

Observed redundancy: A comparison of the number of retinal ganglion cells to the number of neurons in the primary visual cortex shows an increase in the number of sensory neurons in the cortex as compared to the retina. Simoncelli notes that one major argument of critics in that higher up in the sensory pathway there are greater numbers of neurons that handle the processing of sensory information so this should seem to produce redundancy. However, this observation may not be fully relevant because neurons have different neural coding. In his review, Simoncelli notes "cortical neurons tend to have lower firing rates and may use a different form of code as compared to retinal neurons". Cortical Neurons may also have the ability to encode information over longer periods of time than their retinal counterparts. Experiments done in the auditory system have confirmed that redundancy is decreased.

Difficult to test: Estimation of information-theoretic quantities requires enormous amounts of data, and is thus impractical for experimental verification. Additionally, informational estimators are known to be biased. However, some experimental success has occurred.

Need well-defined criteria for what to measure: This criticism illustrates one of the most fundamental issues of the hypothesis. Here, assumptions are made about the definitions of both the inputs and the outputs of the system. The inputs into the visual system are not completely defined, but they are assumed to be encompassed in a collection of natural images. The output must be defined to test the hypothesis, but variability can occur here too based on the choice of which type of neurons to measure, where they are located and what type of responses, such as firing rate or spike times are chosen to be measured.

How to take noise into account: Some argue that experiments that ignore noise, or other physical constraints on the system are too simplistic. However, some researchers have been able to incorporate these elements into their analyses, thus creating more sophisticated systems.

Biomedical Applications

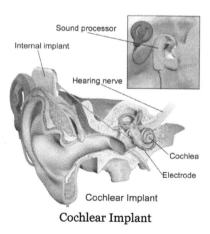

Cochlear Implant

Possible applications of the efficient coding hypothesis include cochlear implant design. These neuroprosthetic devices stimulate the auditory nerve by an electrical impulses which allows some of the hearing to return to people who have hearing impairments or are even deaf. The implants are considered to be successful and efficient and the only ones in use currently. Using frequency-place mappings in the efficient coding algorithm may benefit in the use of cochlear implants in the future. Changes in design based on this hypothesis could increase speech intelligibility in hearing impaired patients. Research using vocoded speech processed by different filters showed that humans had greater accuracy in deciphering the speech when it was processed using an efficient-code filter as opposed to a cochleotropic filter or a linear filter. This shows that efficient coding of noise data offered perceptual benefits and provided the listeners with more information. More research is needed to apply current findings into medically relevant changes to cochlear implant design.

References

- Koch, Christof; Segev, Idan (1999). Methods in neuronal modeling : from ions to networks (2nd ed.). Cambridge, Massachusetts: MIT Press. p. 687. ISBN 0-262-11231-0.

- Gray, Daniel Johnston; Wu, Samuel Miao-Sin (1997). Foundations of cellular neurophysiology (3rd. ed.). Cambridge, Massachusetts [u.a.]: MIT Press. ISBN 9780262100533.

- Hille, Bertil (2001). Ion channels of excitable membranes (3. ed.). Sunderland, Massachusetts: Sinauer. ISBN 9780878933211.

- Nossenson, N.; Magal, N. ,; Messer, H., (2016). ""Detection of stimuli from multi-neuron activity: Empirical study and theoretical implications."". Neurocomputing 174 (2016): 822-837.

4

Tools and Technologies used in Computational Neuroscience

Computational neuroscience is a distinct and complex discipline that uses information technology, neuroscience, psychology, electrical engineering, cognitive science, mathematics and physics. Due to the wide array of subjects that contribute to it, computational neuroscience uses a multitude of interdisciplinary tools and technologies like brain–computer interface, single-unit recording, Bayesian approaches to brain function, neurocomputational speech processing, mind uploading and brain-reading. This chapter provides a comprehensive analysis of the tools and technologies used in this field.

Brain–Computer Interface

A brain–computer interface (BCI), sometimes called a mind-machine interface (MMI), direct neural interface (DNI), or brain–machine interface (BMI), is a direct communication pathway between an enhanced or wired brain and an external device. BCIs are often directed at researching, mapping, assisting, augmenting, or repairing human cognitive or sensory-motor functions.

Research on BCIs began in the 1970s at the University of California, Los Angeles (UCLA) under a grant from the National Science Foundation, followed by a contract from DARPA. The papers published after this research also mark the first appearance of the expression *brain–computer interface* in scientific literature.

The field of BCI research and development has since focused primarily on neuroprosthetics applications that aim at restoring damaged hearing, sight and movement. Thanks to the remarkable cortical plasticity of the brain, signals from implanted prostheses can, after adaptation, be handled by the brain like natural sensor or effector channels. Following years of animal experimentation, the first neuroprosthetic devices implanted in humans appeared in the mid-1990s.

History

The history of brain–computer interfaces (BCIs) starts with Hans Berger's discovery of the electrical activity of the human brain and the development of electroencephalography (EEG). In 1924 Berger was the first to record human brain activity by means of

EEG. Berger was able to identify oscillatory activity, such as Berger's wave or the alpha wave (8–13 Hz), by analyzing EEG traces.

Berger's first recording device was very rudimentary. He inserted silver wires under the scalps of his patients. These were later replaced by silver foils attached to the patients' head by rubber bandages. Berger connected these sensors to a Lippmann capillary electrometer, with disappointing results. However, more sophisticated measuring devices, such as the Siemens double-coil recording galvanometer, which displayed electric voltages as small as one ten thousandth of a volt, led to success.

Berger analyzed the interrelation of alternations in his EEG wave diagrams with brain diseases. EEGs permitted completely new possibilities for the research of human brain activities.

Jacques Vidal coined the term "BCI" and produced the first peer-reviewed publications on this topic. Vidal is widely recognized as the inventor of BCIs in the BCI community, as reflected in numerous peer-reviewed articles reviewing and discussing the field (e.g.,).

Vidal's first BCI relied on visual evoked potentials to allow users to control cursor direction, and visual evoked potentials are still widely used in BCIs (Allison et al., 2010, 2012; Bin et al., 2011; Guger et al., 2012; Kaufmann et al., 2012; Jin et al., 2014; Kapeller et al., 2015).

After his early contributions, Vidal was not active in BCI research, nor BCI events such as conferences, for many years. In 2011, however, he gave a lecture in Graz, Austria, supported by the Future BNCI project, presenting the first BCI, which earned a standing ovation. Vidal was joined by his wife, Laryce Vidal, who previously worked with him at UCLA on his first BCI project. Prof. Vidal will also present a lecture on his early BCI work at the Sixth Annual BCI Meeting, scheduled for May–June 2016 at Asilomar, California.

Versus Neuroprosthetics

Neuroprosthetics is an area of neuroscience concerned with neural prostheses, that is, using artificial devices to replace the function of impaired nervous systems and brain related problems, or of sensory organs. The most widely used neuroprosthetic device is the cochlear implant which, as of December 2010, had been implanted in approximately 220,000 people worldwide. There are also several neuroprosthetic devices that aim to restore vision, including retinal implants.

The difference between BCIs and neuroprosthetics is mostly in how the terms are used: neuroprosthetics typically connect the nervous system to a device, whereas BCIs usually connect the brain (or nervous system) with a computer system. Practical neuroprosthetics can be linked to any part of the nervous system—for example, peripheral nerves—while the term "BCI" usually designates a narrower class of systems which interface with the central nervous system.

The terms are sometimes, however, used interchangeably. Neuroprosthetics and BCIs seek to achieve the same aims, such as restoring sight, hearing, movement, ability to communicate, and even cognitive function. Both use similar experimental methods and surgical techniques.

Animal BCI Research

Several laboratories have managed to record signals from monkey and rat cerebral cortices to operate BCIs to produce movement. Monkeys have navigated computer cursors on screen and commanded robotic arms to perform simple tasks simply by thinking about the task and seeing the visual feedback, but without any motor output. In May 2008 photographs that showed a monkey at the University of Pittsburgh Medical Center operating a robotic arm by thinking were published in a number of well known science journals and magazines. Other research on cats has decoded their neural visual signals.

Early Work

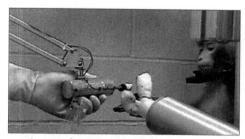

Monkey operating a robotic arm with brain–computer interfacing
(Schwartz lab, University of Pittsburgh)

In 1969 the operant conditioning studies of Fetz and colleagues, at the Regional Primate Research Center and Department of Physiology and Biophysics, University of Washington School of Medicine in Seattle, showed for the first time that monkeys could learn to control the deflection of a biofeedback meter arm with neural activity. Similar work in the 1970s established that monkeys could quickly learn to voluntarily control the firing rates of individual and multiple neurons in the primary motor cortex if they were rewarded for generating appropriate patterns of neural activity.

Studies that developed algorithms to reconstruct movements from motor cortex neurons, which control movement, date back to the 1970s. In the 1980s, Apostolos Georgopoulos at Johns Hopkins University found a mathematical relationship between the electrical responses of single motor cortex neurons in rhesus macaque monkeys and the direction in which they moved their arms (based on a cosine function). He also found that dispersed groups of neurons, in different areas of the monkey's brains, collectively controlled motor commands, but was able to record the firings of neurons in only one area at a time, because of the technical limitations imposed by his equipment.

There has been rapid development in BCIs since the mid-1990s. Several groups have been able to capture complex brain motor cortex signals by recording from neural ensembles (groups of neurons) and using these to control external devices.

Prominent Research Successes

Kennedy and Yang Dan

Phillip Kennedy (who later founded Neural Signals in 1987) and colleagues built the first intracortical brain–computer interface by implanting neurotrophic-cone electrodes into monkeys.

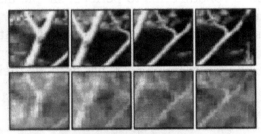

Yang Dan and colleagues' recordings of cat vision using a BCI implanted in the lateral geniculate nucleus
(top row: original image; bottom row: recording)

In 1999, researchers led by Yang Dan at the University of California, Berkeley decoded neuronal firings to reproduce images seen by cats. The team used an array of electrodes embedded in the thalamus (which integrates all of the brain's sensory input) of sharp-eyed cats. Researchers targeted 177 brain cells in the thalamus lateral geniculate nucleus area, which decodes signals from the retina. The cats were shown eight short movies, and their neuron firings were recorded. Using mathematical filters, the researchers decoded the signals to generate movies of what the cats saw and were able to reconstruct recognizable scenes and moving objects. Similar results in humans have since been achieved by researchers in Japan.

Nicolelis

Miguel Nicolelis, a professor at Duke University, in Durham, North Carolina, has been a prominent proponent of using multiple electrodes spread over a greater area of the brain to obtain neuronal signals to drive a BCI.

After conducting initial studies in rats during the 1990s, Nicolelis and his colleagues developed BCIs that decoded brain activity in owl monkeys and used the devices to reproduce monkey movements in robotic arms. Monkeys have advanced reaching and grasping abilities and good hand manipulation skills, making them ideal test subjects for this kind of work.

By 2000 the group succeeded in building a BCI that reproduced owl monkey movements while the monkey operated a joystick or reached for food. The BCI operated in real time and could also control a separate robot remotely over Internet protocol. But the monkeys could not see the arm moving and did not receive any feedback, a so-called open-loop BCI.

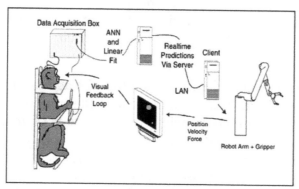

Diagram of the BCI developed by Miguel Nicolelis and colleagues for use on rhesus monkeys

Later experiments by Nicolelis using rhesus monkeys succeeded in closing the feedback loop and reproduced monkey reaching and grasping movements in a robot arm. With their deeply cleft and furrowed brains, rhesus monkeys are considered to be better models for human neurophysiology than owl monkeys. The monkeys were trained to reach and grasp objects on a computer screen by manipulating a joystick while corresponding movements by a robot arm were hidden. The monkeys were later shown the robot directly and learned to control it by viewing its movements. The BCI used velocity predictions to control reaching movements and simultaneously predicted handgripping force. In 2011 O'Doherty and colleagues showed a BCI with sensory feedback with rhesus monkeys. The monkey was brain controlling the position of an avatar arm while receiving sensory feedback through direct intracortical stimulation (ICMS) in the arm representation area of the sensory cortex.

Donoghue, Schwartz and Andersen

Other laboratories which have developed BCIs and algorithms that decode neuron signals include those run by John Donoghue at Brown University, Andrew Schwartz at the University of Pittsburgh and Richard Andersen at Caltech. These researchers have been able to produce working BCIs, even using recorded signals from far fewer neurons than did Nicolelis (15–30 neurons versus 50–200 neurons).

Donoghue's group reported training rhesus monkeys to use a BCI to track visual targets on a computer screen (closed-loop BCI) with or without assistance of a joystick. Schwartz's group created a BCI for three-dimensional tracking in virtual reality and also reproduced BCI control in a robotic arm. The same group also created headlines when they demonstrated that a monkey could feed itself pieces of fruit and marshmallows using a robotic arm controlled by the animal's own brain signals.

Andersen's group used recordings of premovement activity from the posterior parietal cortex in their BCI, including signals created when experimental animals anticipated receiving a reward.

Other Research

In addition to predicting kinematic and kinetic parameters of limb movements, BCIs that predict electromyographic or electrical activity of the muscles of primates are being developed. Such BCIs could be used to restore mobility in paralyzed limbs by electrically stimulating muscles.

Miguel Nicolelis and colleagues demonstrated that the activity of large neural ensembles can predict arm position. This work made possible creation of BCIs that read arm movement intentions and translate them into movements of artificial actuators. Carmena and colleagues programmed the neural coding in a BCI that allowed a monkey to control reaching and grasping movements by a robotic arm. Lebedev and colleagues argued that brain networks reorganize to create a new representation of the robotic appendage in addition to the representation of the animal's own limbs.

The biggest impediment to BCI technology at present is the lack of a sensor modality that provides safe, accurate and robust access to brain signals. It is conceivable or even likely, however, that such a sensor will be developed within the next twenty years. The use of such a sensor should greatly expand the range of communication functions that can be provided using a BCI.

Development and implementation of a BCI system is complex and time consuming. In response to this problem, Gerwin Schalk has been developing a general-purpose system for BCI research, called BCI2000. BCI2000 has been in development since 2000 in a project led by the Brain–Computer Interface R&D Program at the Wadsworth Center of the New York State Department of Health in Albany, New York, United States.

A new 'wireless' approach uses light-gated ion channels such as Channelrhodopsin to control the activity of genetically defined subsets of neurons in vivo. In the context of a simple learning task, illumination of transfected cells in the somatosensory cortex influenced the decision making process of freely moving mice.

The use of BMIs has also led to a deeper understanding of neural networks and the central nervous system. Research has shown that despite the inclination of neuroscientists to believe that neurons have the most effect when working together, single neurons can be conditioned through the use of BMIs to fire at a pattern that allows primates to control motor outputs. The use of BMIs has led to development of the single neuron insufficiency principle which states that even with a well tuned firing rate single neurons can only carry a narrow amount of information and therefore the highest level of accuracy is achieved by recording firings of the collective ensemble. Other principles discovered with the use of BMIs include the neuronal multitasking principle, the neuronal mass principle, the neural degeneracy principle, and the plasticity principle.

BCIs are also proposed to be applied by users without disabilities. A user-centered categorization of BCI approaches by Thorsten O. Zander and Christian Kothe introduces the term passive BCI. Next to active an reactive BCI that are used for directed control, passive BCIs allow for assessing and interpreting changes in the user state during Human-Computer Interaction (HCI). In a secondary, implicit control loop the computer system adapts to its user improving its usability in general.

The BCI Award

The Annual BCI Research Award, endowed with 3,000 USD, is awarded in recognition of outstanding and innovative research in the field of Brain-Computer Interfaces. Each year, a renowned research laboratory is asked to judge the submitted projects and to award the prize. The jury consists of world-leading BCI experts recruited by the awarding laboratory. Following list consists the winners of the BCI Award:

- 2010: Cuntai Guan, Kai Keng Ang, Karen Sui Geok Chua and Beng Ti Ang, (A*STAR, Singapore)

  Motor imagery-based Brain-Computer Interface robotic rehabilitation for stroke.

- 2011: Moritz Grosse-Wentrup and Bernhard Schölkopf, (Max Planck Institute for Intelligent Systems, Germany)

  What are the neuro-physiological causes of performance variations in brain-computer interfacing?

- 2012: Surjo R. Soekadar and Niels Birbaumer, (Applied Neurotechnology Lab, University Hospital Tübingen and Institute of Medical Psychology and Behavioral Neurobiology, Eberhard Karls University, Tübingen, Germany)

  Improving Efficacy of Ipsilesional Brain-Computer Interface Training in Neurorehabilitation of Chronic Stroke.

- 2013: M. C. Dadarlat[a,b], J. E. O'Doherty[a], P. N. Sabes[a,b] ([a]Department of Physiology, Center for Integrative Neuroscience, San Francisco, CA, US, [b]UC Berkeley-UCSF Bioengineering Graduate Program, University of California, San Francisco, CA, US),

  A learning-based approach to artificial sensory feedback: intracortical microstimulation replaces and augments vision.

- 2014: Katsuhiko Hamada, Hiromu Mori, Hiroyuki Shinoda, Tomasz M. Rutkowski, (The University of Tokyo, JP, Life Science Center of TARA, University of Tsukuba, JP, RIKEN Brain Science Institute, JP),

  Airborne Ultrasonic Tactile Display BCI

Human BCI Research

Invasive BCIs

Invasive BCI research has targeted repairing damaged sight and providing new functionality for people with paralysis. Invasive BCIs are implanted directly into the grey matter of the brain during neurosurgery. Because they lie in the grey matter, invasive devices produce the highest quality signals of BCI devices but are prone to scar-tissue build-up, causing the signal to become weaker, or even non-existent, as the body reacts to a foreign object in the brain.

Jens Naumann, a man with acquired blindness, being interviewed
about his vision BCI on CBS's The Early Show

In *vision science*, direct brain implants have been used to treat non-congenital (acquired) blindness. One of the first scientists to produce a working brain interface to restore sight was private researcher William Dobelle.

Dobelle's first prototype was implanted into "Jerry", a man blinded in adulthood, in 1978. A single-array BCI containing 68 electrodes was implanted onto Jerry's visual cortex and succeeded in producing phosphenes, the sensation of seeing light. The system included cameras mounted on glasses to send signals to the implant. Initially, the implant allowed Jerry to see shades of grey in a limited field of vision at a low framerate. This also required him to be hooked up to a mainframe computer, but shrinking electronics and faster computers made his artificial eye more portable and now enable him to perform simple tasks unassisted.

In 2002, Jens Naumann, also blinded in adulthood, became the first in a series of 16 paying patients to receive Dobelle's second generation implant, marking one of the earliest commercial uses of BCIs. The second generation device used a more sophisticated implant enabling better mapping of phosphenes into coherent vision. Phosphenes are spread out across the visual field in what researchers call "the starry-night effect". Immediately after his implant, Jens was able to use his imperfectly restored vision to drive an automobile slowly around the parking area of the research institute. Unfortunately, Dobelle died in 2004 before his processes and developments were documented. Subsequently, when Mr. Naumann and the other patients in the program began having problems with their vision, there was no relief and they eventually lost their "sight" again. Naumann wrote about his experience with Dobelle's work in *Search for Paradise: A Patient's Account of the Artificial Vision Experiment* and has returned to his farm in Southeast Ontario, Canada, to resume his normal activities.

Dummy unit illustrating the design of a BrainGate interface

Movement

BCIs focusing on *motor neuroprosthetics* aim to either restore movement in individuals with paralysis or provide devices to assist them, such as interfaces with computers or robot arms.

Researchers at Emory University in Atlanta, led by Philip Kennedy and Roy Bakay, were first to install a brain implant in a human that produced signals of high enough quality to simulate movement. Their patient, Johnny Ray (1944–2002), suffered from 'locked-in syndrome' after suffering a brain-stem stroke in 1997. Ray's implant was installed in 1998 and he lived long enough to start working with the implant, eventually learning to control a computer cursor; he died in 2002 of a brain aneurysm.

Tetraplegic Matt Nagle became the first person to control an artificial hand using a BCI in 2005 as part of the first nine-month human trial of Cyberkinetics's BrainGate chip-implant. Implanted in Nagle's right precentral gyrus (area of the motor cortex for arm movement), the 96-electrode BrainGate implant allowed Nagle to control a robotic arm by thinking about moving his hand as well as a computer cursor, lights and TV. One year later, professor Jonathan Wolpaw received the prize of the Altran Foundation for Innovation to develop a Brain Computer Interface with electrodes located on the surface of the skull, instead of directly in the brain.

More recently, research teams led by the Braingate group at Brown University and a group led by University of Pittsburgh Medical Center, both in collaborations with the United States Department of Veterans Affairs, have demonstrated further success in direct control of robotic prosthetic limbs with many degrees of freedom using direct connections to arrays of neurons in the motor cortex of patients with tetraplegia.

Partially Invasive BCIs

Partially invasive BCI devices are implanted inside the skull but rest outside the brain rather than within the grey matter. They produce better resolution signals than non-invasive BCIs where the bone tissue of the cranium deflects and deforms signals and have a lower risk of forming scar-tissue in the brain than fully invasive BCIs. There has been preclinical demonstration of intracortical BCIs from the stroke perilesional cortex.

Electrocorticography (ECoG) measures the electrical activity of the brain taken from beneath the skull in a similar way to non-invasive electroencephalography, but the electrodes are embedded in a thin plastic pad that is placed above the cortex, beneath the dura mater. ECoG technologies were first trialled in humans in 2004 by Eric Leuthardt and Daniel Moran from Washington University in St Louis. In a later trial, the researchers enabled a teenage boy to play Space Invaders using his ECoG implant. This research indicates that control is rapid, requires minimal training, and may be an ideal tradeoff with regards to signal fidelity and level of invasiveness.

(Note: these electrodes had not been implanted in the patient with the intention of developing a BCI. The patient had been suffering from severe epilepsy and the electrodes were temporarily implanted to help his physicians localize seizure foci; the BCI researchers simply took advantage of this.)

Signals can be either subdural or epidural, but are not taken from within the brain parenchyma itself. It has not been studied extensively until recently due to the limited access of subjects. Currently, the only manner to acquire the signal for study is through the use of patients requiring invasive monitoring for localization and resection of an epileptogenic focus.

ECoG is a very promising intermediate BCI modality because it has higher spatial resolution, better signal-to-noise ratio, wider frequency range, and less training requirements than scalp-recorded EEG, and at the same time has lower technical difficulty, lower clinical risk, and probably superior long-term stability than intracortical single-neuron recording. This feature profile and recent evidence of the high level of control with minimal training requirements shows potential for real world application for people with motor disabilities.

Light reactive imaging BCI devices are still in the realm of theory. These would involve implanting a laser inside the skull. The laser would be trained on a single neuron and the neuron's reflectance measured by a separate sensor. When the neuron fires, the laser light pattern and wavelengths it reflects would change slightly. This would allow researchers to monitor single neurons but require less contact with tissue and reduce the risk of scar-tissue build-up.

In 2014, a BCI study using near-infrared spectroscopy for "locked-in" patients with amyotrophic lateral sclerosis (ALS) was able to restore some basic ability of the patients to communicate with other people.

Non-invasive BCIs

There have also been experiments in humans using non-invasive neuroimaging technologies as interfaces. The substantial majority of published BCI work involves non-invasive EEG-based BCIs. Noninvasive EEG-based technologies and interfaces have been used for a much broader variety of applications. Although EEG-based interfaces are easy to wear and do not require surgery, they have relatively poor spatial resolution and cannot effectively use higher-frequency signals because the skull dampens signals, dispersing and blurring the electromagnetic waves created by the neurons. EEG-based interfaces also require some time and effort prior to each usage session, whereas non-EEG-based ones, as well as invasive ones require no prior-usage training. Overall, the best BCI for each user depends on numerous factors.

Non EEG-based

Pupil-size Oscillation

In a recent 2016 article, an entirely new communication device and non EEG-based BCI was developed, requiring no visual fixation or ability to move eyes at all, that is based on covert interest in (i.e. without fixing eyes on) chosen letter on a virtual keyboard with letters each having its own (background) circle that is micro-oscillating in brightness in different time transitions, where the letter selection is based on best fit between, on one hand, unintentional pupil-size oscillation pattern, and, on the other hand, the circle-in-background's brightness oscillation pattern. Accuracy is additionally improved by user's mental rehearsing the words 'bright' and 'dark' in synchrony with the brightness transitions of the circle/letter.

EEG-based

Overview

Recordings of brainwaves produced by an electroencephalogram

Electroencephalography (EEG) is the most studied non-invasive interface, mainly due to its fine temporal resolution, ease of use, portability and low set-up cost. The technology is somewhat susceptible to noise however. In the early days of BCI research, another substantial barrier to using EEG as a brain–computer interface was the extensive training required before users can work the technology. For example, in experiments beginning in the mid-1990s, Niels Birbaumer at the University of Tübingen in Germany trained severely paralysed people to self-regulate the *slow cortical potentials* in their EEG to such an extent that these signals could be used as a binary signal to control a computer cursor. (Birbaumer had earlier trained epileptics to prevent impending fits by controlling this low voltage wave.) The experiment saw ten patients trained to move a computer cursor by controlling their brainwaves. The process was slow, requiring more than an hour for patients to write 100 characters with the cursor, while training often took many months. However, the slow cortical potential approach to BCIs has not been used in several years, since other approaches require little or no training, are faster and more accurate, and work for a greater proportion of users.

Another research parameter is the type of oscillatory activity that is measured. Birbaumer's later research with Jonathan Wolpaw at New York State University has focused on developing technology that would allow users to choose the brain signals they found easiest to operate a BCI, including *mu* and *beta* rhythms.

A further parameter is the method of feedback used and this is shown in studies of P300 signals. Patterns of P300 waves are generated involuntarily (stimulus-feedback) when people see something they recognize and may allow BCIs to decode categories of thoughts without training patients first. By contrast, the biofeedback methods described above require learning to control brainwaves so the resulting brain activity can be detected.

Lawrence Farwell and Emanuel Donchin developed an EEG-based brain–computer interface in the 1980s. Their "mental prosthesis" used the P300 brainwave response to allow subjects, including one paralyzed Locked-In syndrome patient, to communicate words, letters and simple commands to a computer and thereby to speak through a speech synthesizer driven by the computer. A number of similar devices have been developed since then. In 2000, for example, research by Jessica Bayliss at the University of Rochester showed that volunteers wearing virtual reality helmets could control elements in a virtual world using their P300 EEG readings, including turning lights on and off and bringing a mock-up car to a stop.

While an EEG based brain-computer interface has been pursued extensively by a number of research labs, recent advancements made by Bin He and his team at the University of Minnesota suggest the potential of an EEG based brain-computer interface to accomplish tasks close to invasive brain-computer interface. Using advanced functional neuroimaging including BOLD functional MRI and EEG source imaging, Bin He and co-workers identified the co-variation and co-localization of electrophysiological and hemodynamic signals induced by motor imagination. Refined by a neuroimaging approach and by a training protocol, Bin He and co-workers demonstrated the ability of a non-invasive EEG based brain-computer interface to control the flight of a virtual helicopter in 3-dimensional space, based upon motor imagination. In June 2013 it was announced that Bin He had developed the technique to enable a remote-control helicopter to be guided through an obstacle course.

In addition to a brain-computer interface based on brain waves, as recorded from scalp EEG electrodes, Bin He and co-workers explored a virtual EEG signal-based brain-computer interface by first solving the EEG inverse problem and then used the resulting virtual EEG for brain-computer interface tasks. Well-controlled studies suggested the merits of such a source analysis based brain-computer interface.

A 2014 study found that severely motor-impaired patients could communicate faster and more reliably with non-invasive EEG BCI, than with any muscle-based communication channel.

Dry Active Electrode Arrays

In the early 1990s Babak Taheri, at University of California, Davis demonstrated the first single and also multichannel dry active electrode arrays using micro-machining. The single channel dry EEG electrode construction and results were published in 1994. The arrayed electrode was also demonstrated to perform well compared to silver/silver chloride electrodes. The device consisted of four sites of sensors with integrated electronics to reduce noise by impedance matching. The advantages of such electrodes are: (1) no electrolyte used, (2) no skin preparation, (3) significantly reduced sensor size, and (4) compatibility with EEG monitoring systems. The active electrode array is an integrated system made of an array of capacitive sensors with local integrated circuitry housed in a package with batteries to power the circuitry. This level of integration was required to achieve the functional performance obtained by the electrode.

The electrode was tested on an electrical test bench and on human subjects in four modalities of EEG activity, namely: (1) spontaneous EEG, (2) sensory event-related potentials, (3) brain stem potentials, and (4) cognitive event-related potentials. The performance of the dry electrode compared favorably with that of the standard wet electrodes in terms of skin preparation, no gel requirements (dry), and higher signal-to-noise ratio.

In 1999 researchers at Case Western Reserve University, in Cleveland, Ohio, led by Hunter Peckham, used 64-electrode EEG skullcap to return limited hand movements to quadriplegic Jim Jatich. As Jatich concentrated on simple but opposite concepts like up and down, his beta-rhythm EEG output was analysed using software to identify patterns in the noise. A basic pattern was identified and used to control a switch: Above average activity was set to on, below average off. As well as enabling Jatich to control a computer cursor the signals were also used to drive the nerve controllers embedded in his hands, restoring some movement.

Prosthesis and Environment Control

Non-invasive BCIs have also been applied to enable brain-control of prosthetic upper and lower extremity devices in people with paralysis. For example, Gert Pfurtscheller of Graz University of Technology and colleagues demonstrated a BCI-controlled functional electrical stimulation system to restore upper extremity movements in a person with tetraplegia due to spinal cord injury. Between 2012 and 2013, researchers at the University of California, Irvine demonstrated for the first time that it is possible to use BCI technology to restore brain-controlled walking after spinal cord injury. In their spinal cord injury research study, a person with paraplegia was able to operate a BCI-robotic gait orthosis to regain basic brain-controlled ambulation. In 2009 Alex Blainey, an independent researcher based in the UK, successfully used the Emotiv EPOC to control a 5 axis robot arm. He then went on to make several demonstration mind controlled wheelchairs and home automation that could be operated by people with limited or no motor control such as those with paraplegia and cerebral palsy.

Other Research

Electronic neural networks have been deployed which shift the learning phase from the user to the computer. Experiments by scientists at the Fraunhofer Society in 2004 using neural networks led to noticeable improvements within 30 minutes of training.

Experiments by Eduardo Miranda, at the University of Plymouth in the UK, has aimed to use EEG recordings of mental activity associated with music to allow the disabled to express themselves musically through an encephalophone. Ramaswamy Palaniappan has pioneered the development of BCI for use in biometrics to identify/authenticate a person. The method has also been suggested for use as PIN generation device (for example in ATM and internet banking transactions. The group which is now at University of Wolverhampton has previously developed analogue cursor control using thoughts.

Researchers at the University of Twente in the Netherlands have been conducting research on using BCIs for non-disabled individuals, proposing that BCIs could improve error handling, task performance, and user experience and that they could broaden the user spectrum. They particularly focused on BCI games, suggesting that BCI games could provide challenge, fantasy and sociality to game players and could, thus, improve player experience.

The first BCI session with 100% accuracy (based on 80 right-hand and 80 left-hand movement imaginations) was recorded in 1998 by Christoph Guger. The BCI system used 27 electrodes overlaying the sensorimotor cortex, weighted the electrodes with Common Spatial Patterns, calculated the running variance and used a linear discriminant analysis.

Research is ongoing into military use of BCIs and since the 1970s DARPA has been funding research on this topic. The current focus of research is user-to-user communication through analysis of neural signals. The project "Silent Talk" aims to detect and analyze the word-specific neural signals, using EEG, which occur before speech is vocalized, and to see if the patterns are generalizable.

DIY and Open Source BCI

In 2001, The OpenEEG Project was initiated by a group of DIY neuroscientists and engineers. The ModularEEG was the primary device created the OpenEEG community; it was a 6-channel signal capture board that cost between $200 and $400 to make at home. The OpenEEG Project marked a significant moment in the emergence of DIY brain-computer interfacing.

In 2010, the Frontier Nerds of NYU's ITP program published a thorough tutorial titled How To Hack Toy EEGs. The tutorial, which stirred the minds of many budding DIY BCI enthusiasts, demonstrated how to create a single channel at-home EEG with an Arduino and a Mattel Mindflex at a very reasonable price. This tutorial amplified the DIY BCI movement.

In 2013, OpenBCI emerged from a DARPA solicitation and subsequent Kickstarter campaign. They created a high-quality, open-source 8-channel EEG acquisition board, known as the 32bit Board, that retailed for under $500. Two years later they created the first 3D-printed EEG Headset, known as the Ultracortex, as well as, a 4-channel EEG acquisition board, known as the Ganglion Board, that retailed for under $100.

In 2015, NeuroTechX was created with the mission of building an international network for neurotechnology. They bring hackers, researchers and enthusiasts all together in many different cities around the world. According to their rapid growth, the DIY neurotech / BCI community was already waiting for such initiative to see light.

MEG and MRI

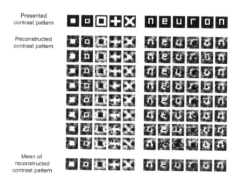

ATR Labs' reconstruction of human vision using fMRI (top row: original image; bottom row: reconstruction from mean of combined readings)

Magnetoencephalography (MEG) and functional magnetic resonance imaging (fMRI) have both been used successfully as non-invasive BCIs. In a widely reported experiment, fMRI allowed two users being scanned to play Pong in real-time by altering their haemodynamic response or brain blood flow through biofeedback techniques.

fMRI measurements of haemodynamic responses in real time have also been used to control robot arms with a seven-second delay between thought and movement.

In 2008 research developed in the Advanced Telecommunications Research (ATR) Computational Neuroscience Laboratories in Kyoto, Japan, allowed the scientists to reconstruct images directly from the brain and display them on a computer in black and white at a resolution of 10x10 pixels. The article announcing these achievements was the cover story of the journal Neuron of 10 December 2008.

In 2011 researchers from UC Berkeley published a study reporting second-by-second reconstruction of videos watched by the study's subjects, from fMRI data. This was achieved by creating a statistical model relating visual patterns in videos shown to the subjects, to the brain activity caused by watching the videos. This model was then used to look up the 100 one-second video segments, in a database of 18 million seconds of random YouTube videos, whose visual patterns most closely matched the brain activity recorded when subjects watched a new video. These 100 one-second video extracts were then combined into a mashed-up image that resembled the video being watched.

Neurogaming

Currently, there is a new field of gaming called Neurogaming, which uses non-invasive BCI in order to improve gameplay so that users can interact with a console without the use of a traditional controller. Some Neurogaming software use a player's brain waves, heart rate, expressions, pupil dilation, and even emotions to complete tasks or affect the mood of the game. For example, game developers at Emotiv have created non-invasive BCI that will determine the mood of a player and adjust music or scenery accordingly. This new form of interaction between player and software will enable a player to have a more realistic gaming experience. Because there will be less disconnect between a player and console, Neurogaming will allow individuals to utilize their "psychological state" and have their reactions transfer to games in real-time.

However, since Neurogaming is still in its first stages, not much is written about the new industry. The first NeuroGaming Conference was held in San Francisco on May 1–2, 2013.

BCI Control Strategies in Neurogaming

Motor Imagery

Motor imagery involves the imagination of the movement of various body parts resulting in sensorimotor cortex activation,which modulates sensorimotor oscillations in the EEG. This can be detected by the BCI to infer a user's intent. Motor imagery typically requires a number of sessions of training before acceptable control of the BCI is acquired. These training sessions may take a number of hours over several days before users can consistently employ the technique with acceptable levels of precision. Regardless of the duration of the training session, users are unable to master the control scheme. This results in very slow pace of the gameplay. Advance machine learning methods were recently developed to compute a subject-specific model for detecting the performance of motor imagery. The top performing algorithm from BCI Competition IV (http://www.bbci.de/competition/iv/) dataset 2 for motor imagery is the Filter Bank Common Spatial Pattern, developed by Ang et al. from A*STAR, Singapore).

Bio/Neurofeedback for Passive BCI Designs

Biofeedback is used to monitor a subject's mental relaxation. In some cases, biofeedback does not monitor electroencephalography (EEG), but instead bodily parameters such as electromyography(EMG), galvanic skin resistance (GSR), and heart rate variability (HRV).Many biofeedback systems are used to treat certain disorders such as attention deficit hyperactivity disorder (ADHD), sleep problems in children, teeth grinding, and chronic pain. EEG biofeedback systems typically monitor four different bands (theta: 4–7 Hz, alpha:8–12 Hz, SMR: 12–15 Hz, beta: 15–18 Hz) and challenge the subject to control them. Passive BCI involves using BCI to enrich human–machine interaction with implicit information on the actual user's state, for example, simulations to detect when users intend to push brakes during an emergency car stopping procedure. Game developers using passive BCIs need to acknowledge that through repetition of game levels the user's cognitive state will change or adapt. Within the first play of a level, the user will react to things differently from during the second play: for example, the user will be less surprised at an event in the game if he/she is expecting it.

Visual Evoked Potential (VEP)

A VEP is an electrical potential recorded after a subject is presented with a type of visual stimuli. There are several types of VEPs.

Steady-state visually evoked potentials (SSVEPs) use potentials generated by exciting the retina, using visual stimuli modulated at certain frequencies. SSVEP's stimuli are often formed from alternating checkerboard patterns and at times simply use flashing images . The frequency of the phase reversal of the stimulus used can be clearly distinguished in the spectrum of an EEG; this makes detection of SSVEP stimuli relatively easy . SSVEP has proved to be successful within many BCI systems . This is due to several factors, the signal elicited is measurable in as large a population as the transient VEP and blink movement and electro cardiographic artefacts do not affect the frequencies monitored. In addition, the SSVEP signal is exceptionally robust; the topographic organization of the primary visual cortex is such that a broader area obtains afferents from the central or fovial region of the visual field .SSVEP does have several problems however. As SSVEPs use flashing stimuli to infer a user's intent, the user must gaze at one of the flashing or iterating symbols in order to interact with the system. It is, therefore, likely that the symbols could become irritating and uncomfortable to use during longer play sessions, which can often last more than an hour which may not be an ideal gameplay.

Another type of VEP used with applications is the P300 potential. The P300 event-related potential is a positive peak in the EEG that occurs at roughly 300 ms after the appearance of a target stimulus (a stimulus for which the user is waiting or seeking) or oddball stimuli . The P300 amplitude decreases as the target stimuli and the ignored stimuli grow more similar.The P300 is thought to be related to a higher level attention process or an orienting response Using P300 as a control scheme has the advantage of the participant only having to attend limited training sessions. The first application to use the P300 model was the P300 matrix . Within this system, a subject would choose a letter from a grid of 6 by 6 letters and numbers. The rows and columns of the grid flashed sequentially and every time the selected "choice letter" was illuminated the user's P300 was (potentially) elicited. However, the communication process, at approximately 17 characters per minute, was quite slow. The P300 is a BCI that offers a discrete selection rather than a continuous control mechanism. The advantage of P300 use within games is that the player does not have to teach himself/herself how to use a completely new control system and so only has to undertake short training instances, to learn the gameplay mechanics and basic use of the BCI paradigm.

Synthetic Telepathy/Silent Communication

In a $6.3 million Army initiative to invent devices for telepathic communication, Gerwin Schalk, underwritten in a $2.2 million grant, found that it is possible to use ECoG signals to discriminate the vowels and consonants embedded in spoken and in imagined words. The results shed light on the distinct mechanisms associated with production of vowels and consonants, and could provide the basis for brain-based communication using imagined speech.

Research into synthetic telepathy using subvocalization is taking place at the University of California, Irvine under lead scientist Mike D'Zmura. The first such communication took place in the 1960s using EEG to create Morse code using brain alpha waves. Using EEG to communicate imagined speech is less accurate than the invasive method of placing an electrode between the skull and the brain. On February 27, 2013 the group of Miguel Nicolelis at Duke University and IINN-ELS successfully connected the brains of two rats with electronic interfaces that allowed them to directly share information, in the first-ever direct brain-to-brain interface.

On 3 September 2014, scientists reported that direct communication between human brains was possible over extended distances through Internet transmission of EEG signals.

In March and in May 2014 a study conducted by Dipartimento di Psicologia Generale - Università di Padova, EVANLAB - Firenze, LiquidWeb s.r.l. company and Dipartimento di Ingegneria e Architettura - Università di Trieste, reports confirmatory results analyzing the EEG activity of two human partners spatially separated approximately 190 km apart when one member of the pair receives the stimulation and the second one is connected only mentally with the first.

Cell-culture BCIs

Researchers have built devices to interface with neural cells and entire neural networks in cultures outside animals. As well as furthering research on animal implantable devices, experiments on cultured neural tissue have focused on building problem-solving networks, constructing basic computers and manipulating robotic devices. Research into techniques for stimulating and recording from individual neurons grown on semiconductor chips is sometimes referred to as neuroelectronics or neurochips.

The world's first Neurochip, developed by Caltech researchers Jerome Pine and Michael Maher

Development of the first working neurochip was claimed by a Caltech team led by Jerome Pine and Michael Maher in 1997. The Caltech chip had room for 16 neurons.

In 2003 a team led by Theodore Berger, at the University of Southern California, started work on a neurochip designed to function as an artificial or prosthetic hippocampus. The neurochip was designed to function in rat brains and was intended as a prototype for the eventual development of higher-brain prosthesis. The hippocampus was chosen because it is thought to be the most ordered and structured part of the brain and is the most studied area. Its function is to encode experiences for storage as long-term memories elsewhere in the brain.

In 2004 Thomas DeMarse at the University of Florida used a culture of 25,000 neurons taken from a rat's brain to fly a F-22 fighter jet aircraft simulator. After collection, the cortical neurons were cultured in a petri dish and rapidly began to reconnect themselves to form a living neural network. The cells were arranged over a grid of 60 electrodes and used to control the pitch and yaw functions of the simulator. The study's focus was on understanding how the human brain performs and learns computational tasks at a cellular level.

Ethical Considerations

Important ethical, legal and societal issues related to brain-computer interfacing are:

- conceptual issues (researchers disagree over what is and what is not a brain-computer interface),
- obtaining informed consent from people who have difficulty communicating,
- risk/benefit analysis,
- shared responsibility of BCI teams (e.g. how to ensure that responsible group decisions can be made),
- the consequences of BCI technology for the quality of life of patients and their families,
- side-effects (e.g. neurofeedback of sensorimotor rhythm training is reported to affect sleep quality),
- personal responsibility and its possible constraints (e.g. who is responsible for erroneous actions with a neuroprosthesis),
- issues concerning personality and personhood and its possible alteration,
- therapeutic applications and their possible exceedance,
- questions of research ethics that arise when progressing from animal experimentation to application in human subjects,
- mind-reading and privacy,
- mind-control,
- use of the technology in advanced interrogation techniques by governmental authorities,
- selective enhancement and social stratification.
- communication to the media.

Clausen stated in 2009 that "BCIs pose ethical challenges, but these are conceptually similar to those that bioethicists have addressed for other realms of therapy". Moreover, he suggests that bioethics is well-prepared to deal with the issues that arise with BCI technologies. Haselager and colleagues pointed out that expectations of BCI efficacy and value play a great role in ethical analysis and the way BCI scientists should approach media. Furthermore, standard protocols can be implemented to ensure ethically sound informed-consent procedures with locked-in patients.

Researchers are well aware that sound ethical guidelines, appropriately moderated enthusiasm in media coverage and education about BCI systems will be of utmost importance for the societal acceptance of this technology. Thus, recently more effort is made inside the BCI community to create consensus on ethical guidelines for BCI research, development and dissemination.

Clinical and Research-grade BCI-based Interfaces

Some companies have been producing high-end systems that have been widely used in established BCI labs for several years. These systems typically entail more channels than the low-cost systems below, with much higher signal quality and robustness in real-world settings. Some systems from new companies have been gaining attention for new BCI applications for new user groups, such as persons with stroke or coma.

- In 2011, Nuamps EEG from www.neuroscan.com was used to study the extent of detectable brain signals from stroke patients who performed motor imagery using BCI in a large clinical trial, and the results showed that majority of the patients (87%) could use the BCI.

- In March 2012 g.tec introduced the intendiX-SPELLER, the first commercially available BCI system for home use which can be used to control computer games and apps. It can detect different brain signals with an accuracy of 99%. has hosted several workshop tours to demonstrate the intendiX system and other hardware and software to the public, such as a workshop tour of the US West Coast during September 2012.

- A German-based company called BrainProducts makes systems that are widely used within established BCI labs.

- In 2012 an Italian startup company, Liquidweb s.r.l., released "Braincontrol", a first prototype of an AAC BCI-based, designed for patients in locked-in state. It was validated from 2012 and 2014 with the involvement of LIS and CLIS patients. In 2014 the company introduced the commercial version of the product, with the CE mark class I as medical device.

Low-cost BCI-based Interfaces

Recently a number of companies have scaled back medical grade EEG technology (and in one case, NeuroSky, rebuilt the technology from the ground up to create inexpensive BCIs. This technology has been built into toys and gaming devices; some of these toys have been extremely commercially successful like the NeuroSky and Mattel MindFlex.

- In 2006 Sony patented a neural interface system allowing radio waves to affect signals in the neural cortex.

- In 2007 NeuroSky released the first affordable consumer based EEG along with the game NeuroBoy. This was also the first large scale EEG device to use dry sensor technology.

- In 2008 OCZ Technology developed a device for use in video games relying primarily on electromyography.

- In 2008 the Final Fantasy developer Square Enix announced that it was partnering with NeuroSky to create a game, Judecca.

- In 2009 Mattel partnered with NeuroSky to release the Mindflex, a game that used an EEG to steer a ball through an obstacle course. By far the best selling consumer based EEG to date.

- In 2009 Uncle Milton Industries partnered with NeuroSky to release the Star Wars Force Trainer, a game designed to create the illusion of possessing The Force.

- In 2009 Emotiv released the EPOC, a 14 channel EEG device that can read 4 mental states, 13 conscious states, facial expressions, and head movements. The EPOC is the first commercial BCI to use dry sensor technology, which can be dampened with a saline solution for a better connection.

- In November 2011 Time Magazine selected "necomimi" produced by Neurowear as one of the best inventions of the year. The company announced that it expected to launch a consumer version of the garment, consisting of cat-like ears controlled by a brain-wave reader produced by NeuroSky, in spring 2012.

- In February 2014 They Shall Walk (a nonprofit organization fixed on constructing exoskeletons, dubbed LIFESUITs, for paraplegics and quadriplegics) began a partnership with James W. Shakarji on the development of a wireless BCI.

Future Directions for BCIs

A consortium consisting of 12 European partners has completed a roadmap to support the European Commission in their funding decisions for the new framework program Horizon 2020. The project, which was funded by the European Commission, started in November 2013 and ended in April 2015. The roadmap is now complete, and can be downloaded on the project's webpage. A 2015 publication describes some of the analyses and achievements of this project, as well as the emerging Brain-Computer Interface Society. For example, this article reviewed work within this project that further defined BCIs and applications, explored recent trends, discussed ethical issues, and evaluated different directions for new BCIs. As the article notes, their new roadmap generally extends and supports the recommendations from the Future BNCI project, which conveys some enthusiasm for emerging BCI directions.

In addition to, other recent publications have explored the most promising future BCI directions for new groups of disabled users (e.g.,). Some prominent examples are summarized below.

Disorders of Consciousness (DOC)

Some persons have a disorder of consciousness (DOC). This state is defined to include persons with coma, as well as persons in a vegetative state (VS) or minimally conscious state (MCS). New BCI research seeks to help persons with DOC in different ways. A key initial goal is to identify patients who are able to perform basic cognitive tasks, which would of course lead to a change in their diagnosis. That is, some persons who are diagnosed with DOC may in fact be able to process information and make important life decisions (such as whether to seek therapy, where to live, and their views on end-of-life decisions regarding them). Very sadly, some persons who are diagnosed with DOC die as a result of end-of-life decisions, which may be made by family members who sincerely feel this is in the patient's best interests. Given the new prospect of allowing these patients to provide their views on this decision, there would seem to be a strong ethical pressure to develop this research direction to guarantee that DOC patients are given an opportunity to decide whether they want to live.

These and other articles describe new challenges and solutions to use BCI technology to help persons with DOC. One major challenge is that these patients cannot use BCIs based on vision. Hence, new tools rely on auditory and/or vibrotactile stimuli. Patients may wear headphones and/or vibrotactile stimulators placed on the wrists, neck, leg, and/or other locations. Another challenge is that patients may fade in and out of consciousness, and can only communicate at certain times. This may indeed be a cause of mistaken diagnosis. Some patients may only be able to respond to physicians' requests during a few hours per day (which might not be predictable ahead of time) and thus may have been unresponsive during diagnosis. Therefore, new methods rely on tools that are easy to use in field settings, even without expert help, so family members and other persons without any medical or technical background can still use them. This reduces the cost, time, need for expertise, and other burdens with DOC assessment. Automated tools can ask simple questions that patients can easily answer, such as "Is your father named George?" or "Were you born in the USA?" Automated instructions inform patients that they may convey yes or no by (for example) focusing their attention on stimuli on the right vs. left wrist. This focused attention produces reliable changes in EEG patterns that can help determine that the patient is able to communicate. The results could be presented to physicians and therapists, which could lead to a revised diagnosis and therapy. In addition, these patients could then be provided with BCI-based communication tools that could help them convey basic needs, adjust bed position and HVAC (heating, ventilation, and air conditioning), and otherwise empower them to make major life decisions and communicate.

This research effort was supported in part by different EU-funded projects, such as the DECODER project led by Prof. Andrea Kuebler at the University of Wuerzburg. This project contributed to the first BCI system developed for DOC assessment and communication, called mindBEAGLE. This system is designed to help non-expert users work with DOC patients, but is not intended to replace medical staff. An EU-funded project scheduled to begin in 2015 called ComAlert will conduct further research and development to improve DOC prediction, assessment, rehabilitation, and communication, called "PARC" in that project. Another project funded by the National Science Foundation is led by Profs. Dean Krusienski and Chang Nam. This project provides for improved vibrotactile systems, advanced signal analysis, and other improvements for DOC assessment and communication.

Functional Brain Mapping

Each year, about 400,000 people undergo brain mapping during neurosurgery. This procedure is often required for people with tumors or epilepsy that do not respond to medication. During this procedure, electrodes are placed on the brain to precisely identify the locations of structures and functional areas. Patients may be awake during neurosurgery and asked to perform certain tasks, such as moving fingers or repeating words. This is necessary so that surgeons can remove only the desired tissue while sparing other regions, such as critical movement or language regions. Removing too much brain tissue can cause permanent damage, while removing too little tissue can leave the underlying condition untreated and require additional neurosurgery. Thus, there is a strong need to improve both methods and systems to map the brain as effectively as possible.

In several recent publications, BCI research experts and medical doctors have collaborated to explore new ways to use BCI technology to improve neurosurgical mapping. This work focuses largely on high gamma activity, which is difficult to detect with non-invasive means. Results have led to improved methods for identifying key areas for movement, language, and other functions. A recent article addressed advances in functional brain mapping and summarizes a workshop.

BCI Society

Many people within the BCI community have been working toward an official Brain-Computer Interface Society over the last few years. At the Fifth International BCI Meeting in Asilomar, CA in 2013, a plenary session of the attendees unanimously voted in favor of forming this Society. Since then, several people have been active developing bylaws, articles of incorporation, official statements, a membership infrastructure, official website, and other details. The Board consists of many of the most established people in BCI research, including three officers: Prof. Jonathan Wolpaw (President), Prof. Nick Ramsey (Vice-President), and Dr. Christoph Guger (Treasurer).

Single-unit Recording

In neuroscience, single-unit recordings provide a method of measuring the electro-physiological responses of single neurons using a microelectrode system. When a neuron generates an action potential, the signal propagates down the neuron as a current which flows in and out of the cell through excitable membrane regions in the soma and axon. A microelectrode is inserted into the brain, where it can record the rate of change in voltage with respect to time. These microelectrodes must be fine-tipped, high-impedance conductors; they are primarily glass micro-pipettes or metal microelectrodes made of platinum or tungsten. Microelectrodes can be carefully placed within (or close to) the cell membrane, allowing the ability to record intracellularly or extracellularly.

Single-unit recordings are widely used in cognitive science, where it permits the analysis of human cognition and cortical mapping. This information can then be applied to brain machine interface (BMI) technologies for brain control of external devices.

Overview

There are many techniques available to record brain activity—including electroencephalography (EEG), magnetoencephalography (MEG), and functional magnetic resonance imaging (fMRI)—but these do not allow for single-neuron resolution. Neurons are the basic functional units in the brain; they transmit information through the body using electrical signals called action potentials. Currently, single-unit recordings provide the most precise recordings from single neurons. A single unit is defined as a single, firing neuron whose spike potentials are distinctly isolated by a recording microelectrode.

The ability to record signals from neurons is centered around the electric current flow through the neuron. As an action potential propagates through the cell, the electric current flows in and out of the soma and axons at excitable membrane regions. This current creates a measurable, changing voltage potential within (and outside) the cell. This allows for two basic types of single-unit recordings. Intracellular single-unit recordings occur within the neuron and measure the voltage change (with respect to time) across the membrane during action potentials. This outputs as a trace with information on membrane resting potential, postsynaptic potentials and spikes through the soma (or axon). Alternatively, when the microelectrode is close to the cell surface extracellular recordings measure the voltage change (with respect to time) outside the cell, giving only spike information. Different types of microelectrodes can be used for single-unit recordings; they are typically high-impedance, fine-tipped and conductive. Fine tips allow for easy penetration without extensive damage to the cell, but they also correlate with high impedance. Additionally, electrical and/or ionic conductivity allow for recordings from both non-polarizable and polarizable electrodes. The two primary classes of electrodes are glass micropipettes and metal electrodes. Electrolyte-filled glass micropipettes are mainly used for intracellular single-unit recordings; metal electrodes (commonly made of stainless steel, platinum, tungsten or iridium) and used for both types of recordings.

Single-unit recordings have provided tools to explore the brain and apply this knowledge to current technologies. Cognitive scientists have used single-unit recordings in the brains of animals and humans to study behaviors and functions. Electrodes can also be inserted into the brain of epileptic patients to determine the position of epileptic foci. More recently, single-unit recordings have been used in brain machine interfaces (BMI). BMIs record brain signals and decode an intended response, which then controls the movement of an external device (such as a computer cursor or prosthetic limb).

History

The ability to record from single units started with the discovery that the nervous system has electrical properties. Since then, single unit recordings have become an important method for understanding mechanisms and functions of the nervous system. Over the years, single unit recording continued to provide insight on topographical mapping of the cortex. Eventual development of microelectrode arrays allowed recording from multiple units at a time.

- 1790s: The first evidence of electrical activity in the nervous system was observed by Luigi Galvani in the 1790s with his studies on dissected frogs. He discovered that you can induce a dead frog leg to twitch with a spark.

- 1888: Santiago Ramón y Cajal, a Spanish neuroscientist, revolutionized neuroscience with his neuron theory, describing the structure of the nervous system and presence of basic functional units— neurons. He won the Nobel Prize in Physiology or Medicine for this work in 1906.

- 1928: The first account of being able to record from the nervous system was by Edgar Adrian in his 1928 publication "The Basis of Sensation". In this, he describes his recordings of electrical discharges in single nerve fibers using a Lippmann electrometer. He won the Nobel Prize in 1932 for his work revealing the function of neurons.

- 1940: Renshaw, Forbes & Morrison performed original studies recording discharge of pyramidal cells in the hippocampus using glass microelectrodes in cats.

- 1950: Woldring and Dirken report the ability to obtain spike activity from the surface of the cerebral cortex with platinum wires.

- 1952: Li and Jasper applied the Renshaw, Forbes, & Morrison method to study electrical activity in the cerebral cortex of a cat. Hodgkin–Huxley model was revealed, where they used a squid giant axon to determine the exact mechanism of action potentials.

- 1953: Iridium microelectrodes developed for recording.

- 1957: John Eccles used intracellular single-unit recording to study synaptic mechanisms in motoneurons (for which he won the Nobel Prize in 1963).

- 1958: Stainless steel microelectrodes developed for recording.

- 1959: Studies by David H. Hubel and Torsten Wiesel. They used single neuron recordings to map the visual cortex in unanesthesized, unrestrained cats using tungsten electrodes. This work won them the Nobel Prize in 1981 for information processing in the visual system.

- 1960: Glass-insulated platinum microelectrodes developed for recording.

- 1967: The first record of multi-electrode arrays for recording was published by Marg and Adams. They applied this method to record many units at a single time in a single patient for diagnostic and therapeutic brain surgery.

- 1978: Schmidt et al. implanted chronic recording micro-cortical electrodes into the cortex of monkeys and showed that they could teach them to control neuronal firing rates, a key step to the possibility of recording neuronal signals and using them for BMIs.

- 1981: Kruger and Bach assemble 30 individual microelectrodes in a 5x6 configuration and implant the electrodes for simultaneous recording of multiple units.

- 1992: Development of the "Utah Intracortical Electrode Array (UIEA), a multiple-electrode array which can access the columnar structure of the cerebral cortex for neurophysiological or neuroprosthetic applications".

- 1994: The Michigan array, a silicon planar electrode with multiple recording sites, was developed. NeuroNexus, a private neurotechnology company, is formed based on this technology.

- 1998: A key breakthrough for BMIs was achieved by Kennedy and Bakay with development of neurotrophic electrodes. In patients with amyotrophic lateral sclerosis (ALS), a neurological condition affecting the ability to control voluntary movement, they were able to successfully record action potentials using microelectrode arrays to control a computer cursor.

Electrophysiology

The basis of single-unit recordings relies on the ability to record electrical signals from neurons.

Neuronal Potentials and Electrodes

When a microelectrode is inserted into an aqueous ionic solution, there is a tendency for cations and anions to react with the electrode creating an electrode-electrolyte interface. The forming of this layer has been termed the Helmholtz layer. A charge distribution occurs across the electrode, which creates a potential which can be measured against a reference electrode. The method of neuronal potential recording is dependent on the type of electrode used. Non-polarizable electrodes are reversible (ions in the solution are charged and discharged). This creates a current flowing through the electrode, allowing for voltage measurement through the electrode with respect to time. Typically, non-polarizable electrodes are glass micropipettes filled with an ionic solution or metal. Alternatively, ideal polarized electrodes do not have the transformation of ions; these are typically metal electrodes. Instead, the ions and electrons at the surface of the metal become polarized with respect to the potential of the solution. The charges orient at the interface to create an electric double layer; the metal then acts like a capacitor. The change in capacitance with respect to time can be measured and converted to voltage using a bridge circuit. Using this technique, when neurons fire an action potential they create changes in potential fields that can be recorded using microelectrodes.

Intracellularly, the electrodes directly record the firing of action, resting and postsynaptic potentials. When a neuron fires, current flows in and out through excitable regions in the axons and cell body of the neuron. This creates potential fields around the neuron. An electrode near a neuron can detect these extracellular potential fields, creating a spike.

Experimental Setup

The basic equipment needed to record single units is microelectrodes, amplifiers, micromanipulators and recording devices. The type of microelectrode used will depend on the application. The high resistance of these electrodes creates a problem during signal amplification. If it were connected to a conventional amplifier with low input resistance, there would be a large potential drop across the microelectrode and the amplifier would only measure a small portion of the true potential. To solve this problem, a cathode follower amplifier must be used as an impedance matching device to collect the voltage and feed it to a conventional amplifier. To record from a single neuron, micromanipulators must be used to precisely insert an electrode into the brain. This is especially important for intracellular single-unit recording.

Finally, the signals must be exported to a recording device. After amplification, signals are filtered with various techniques. They can be recorded by an oscilloscope and camera, but more modern techniques convert the signal with an analog-to-digital converter and output to a computer to be saved. Data-processing techniques can allow for separation and analysis of single units.

Types of Microelectrodes

There are two main types of microelectrodes used for single-unit recordings: glass micropipettes and metal electrodes. Both are high-impedance electrodes, but glass micropipettes are highly resistive and metal electrodes have frequency-dependent impedance. Glass micropipettes are ideal for resting- and action-potential measurement, while metal electrodes are best used for extracellular spike measurements. Each type has different properties and limitations, which can be beneficial in specific applications.

Glass Micropipettes

Glass micropipettes are filled with an ionic solution to make them conductive; a silver-silver chloride (Ag-AgCl) electrode is dipped into the filling solution as an electrical terminal. Ideally, the ionic solutions should have ions similar to ionic species around the electrode; the concentration inside the electrode and surrounding fluid should be the same. Additionally, the diffusive characteristics of the different ions within the electrode should be similar. The ion must also be able to "provide current carrying capacity adequate for the needs of the experiment". And importantly, it must not cause biological changes in the cell it is recording from. Ag-AgCl electrodes are primarily used with a potassium chloride (KCl) solution. With Ag-AgCl electrodes, ions react with it to produce electrical gradients at the interface, creating a voltage change with respect to time. Electrically, glass microelectrode tips have high resistance and high capacitance. They have a tip size of approximately 0.5-1.5 μm with a resistance of about 10-50 MΩ. The small tips make it easy to penetrate the cell membrane with minimal damage for intracellular recordings. Micropipettes are ideal for measurement of resting membrane potentials and with some adjustments can record action potentials. There are some issues to consider when using glass micropipettes. To offset high resistance in glass micropipettes, a cathode follower must be used as the first-stage amplifier. Additionally, high capacitance develops across the glass and conducting solution which can attenuate high-frequency responses. There is also electrical interference inherent in these electrodes and amplifiers.

Metal

Metal electrodes are made of various types of metals, typically silicon, platinum, and tungsten. They "resemble a leaky electrolytic capacitor, having a very high low-frequency impedance and low high-frequency impedance". They are more suitable for measurement of extracellular action potentials, although glass micropipettes can also be used. Metal electrodes are beneficial in some cases because they have high signal-to-noise due to lower impedance for the frequency range of spike signals. They also have better mechanical stiffness for puncturing through brain tissue. Lastly, they are more easily fabricated into different tip shapes and sizes at large quantities. Platinum electrodes are platinum black plated and insulated with glass. "They normally give stable recordings, a high signal-to-noise ratio, good isolation, and they are quite rugged in the usual tip sizes". The only limitation is that the tips are very fine and fragile. Silicon electrodes are alloy electrodes doped with silicon and an insulating glass cover layer. Silicon technology provides better mechanical stiffness and is a good supporting carrier to allow for multiple recording sites on a single electrode. Tungsten electrodes are very rugged and provide very stable recordings. This allows manufacturing of tungsten electrodes with very small tips to isolate high-frequencies. Tungsten, however, is very noisy at low frequencies. In mammalian nervous system where there are fast signals, noise can be removed with a high-pass filter. Slow signals are lost if filtered so tungsten is not a good choice for recording these signals.

Types of Single-unit Recordings

Single unit recordings can be done either intracellularly or extracellularly. While extracellular recordings can only give spike information, intracellular single unit recordings can give information on resting potentials and postsynaptic potentials. The use of either technique depends on the specific application and what information is desired.

Intracellular

Intracellular single unit recordings require electrodes be inserted through the cell membrane to record from within the cell. Glass micropipettes or metal electrodes may be used for intracellular single unit recordings, but glass micropipettes are preferred because their high input resistance allows more precise recordings for measurement of resting potentials. Additionally, very fine glass tip micropipettes are much better at successfully penetrating and retaining neurons. Intracellular single unit recordings provide much more information on single neuron discharges. They can give information on steady and resting membrane voltage, postsynaptic potentials, and spikes (action potentials) from both the axon and cell body. Limitations of intracellular recording are that one can only record from cell bodies of, usually, the largest cells. There is little information obtained on neural information transfer from further dendrites or axons. Recordings from small neurons are quite difficult and usually must be supported with extracellular single unit recordings.

Extracellular

Extracellular single unit recordings are more suitable for measuring extracellular action potentials. They are measured using either glass micropipettes or metal electrodes that are placed close to the neuron. Extracellular recordings can easily measure spike discharge from a neuron with any suitably small electrode. Single neurons can also be isolated and recorded for longer periods of time with no worry of damage to the cells. This makes it much easier to obtain these signals in an awake and moving animal. Limitations of extracellular recordings are that signal detection is a primary concern and it is unable to give information on postsynaptic potentials or resting membrane potentials.

Combined Recordings

More recently, efforts have been made to obtain extracellular and intracellular recordings simultaneously. This involves careful placement of extracellular and intracellular electrodes in a single neuron. The primary use for this is to provide a better understanding of the relationship between intracellular action potentials and extracellular spike recordings.

Applications

Single-unit recordings have allowed the ability to monitor single-neuron activity. This has allowed researchers to discover the role of different parts of the brain in function and behavior. More recently, recording from single neurons can be used to engineer "mind-controlled" devices.

Cognitive Science

Noninvasive tools to study the CNS have been developed to provide structural and functional information, but they do not provide very high resolution. To offset this problem invasive recording methods have been used. Single unit recording methods give high spatial and temporal resolution to allow for information assessing the relationship between brain structure, function, and behavior. By looking at brain activity at the neuron level, researchers can link brain activity to behavior and create neuronal maps describing flow of information through the brain. For example, Boraud et al. report the use of single unit recordings to determine the structural organization of the basal ganglia in patients with Parkinson's disease. Evoked potentials provide a method to couple behavior to brain function. By stimulating different responses, one can visualize what portion of the brain is activated. This method has been used to explore cognitive functions such as perception, memory, language, emotions, and motor control.

Brain-machine Interfaces

Brain-machine interfaces (BMIs) have been developed within the last 20 years. By recording single unit potentials, these devices can decode signals through a computer and output this signal for control of an external device such as a computer cursor or prosthetic limb. BMIs have the potential to restore function in patients with paralysis or neurological disease. This technology has potential to reach a wide variety of patients but is not yet available clinically due to lack of reliability in recording signals over time. The primary hypothesis regarding this failure is that the chronic inflammatory response around the electrode causes neurodegeneration that reduces the number of neurons it is able to record from (Nicolelis, 2001). In 2004, the BrainGate pilot clinical trial was initiated to "test the safety and feasibility of a neural interface system based on an intracortical 100-electrode silicon recording array". This initiative has been successful in advancement of BCIs and in 2011, published data showing long term computer control in a patient with tetraplegia (Simeral, 2011).

Bayesian Approaches to Brain Function

Bayesian approaches to brain function investigate the capacity of the nervous system to operate in situations of uncertainty in a fashion that is close to the optimal prescribed by Bayesian statistics. This term is used in behavioural sciences and neuroscience and studies associated with this term often strive to explain the brain's cognitive abilities based on statistical principles. It is frequently assumed that the nervous system maintains internal probabilistic models that are updated by neural processing of sensory information using methods approximating those of Bayesian probability.

Origins

This field of study has its historical roots in numerous disciplines including machine learning, experimental psychology and Bayesian statistics. As early as the 1860s, with the work of Hermann Helmholtz in experimental psychology the brain's ability to extract perceptual information from sensory data was modeled in terms of probabilistic estimation. The basic idea is that the nervous system needs to organize sensory data into an accurate internal model of the outside world.

Bayesian probability has been developed by many important contributors. Pierre-Simon Laplace, Thomas Bayes, Harold Jeffreys, Richard Cox and Edwin Jaynes developed mathematical techniques and procedures for treating probability as the degree of plausibility that could be assigned to a given supposition or hypothesis based on the available evidence. In 1988 E.T. Jaynes presented a framework for using Bayesian Probability to model mental processes. It was thus realized early on that the Bayesian statistical framework holds the potential to lead to insights into the function of the nervous system.

This idea was taken up in research on unsupervised learning, in particular the Analysis by Synthesis approach, branches of machine learning. In 1983 Geoffrey Hinton and colleagues proposed the brain could be seen as a machine making decisions based on the uncertainties of the outside world. During the 1990s researchers including Peter Dayan, Geoffrey Hinton and Richard Zemel proposed that the brain represents knowledge of the world in terms of probabilities and made specific proposals for tractable neural processes that could manifest such a Helmholtz Machine.

Psychophysics

A wide range of studies interpret the results of psychophysical experiments in light of Bayesian perceptual models. Many aspects of human perceptual and motor behavior can be modeled with Bayesian statistics. This approach, with its emphasis on behavioral outcomes as the ultimate expressions of neural information processing, is also known for modeling sensory and motor decisions using Bayesian decision theory. Examples are the work of Landy, Jacobs, Jordan, Knill, Kording and Wolpert, and Goldreich.

Neural Coding

Many theoretical studies ask how the nervous system could implement Bayesian algorithms. Examples are the work of Pouget, Zemel, Deneve, Latham, Hinton and Dayan. George and Hawkins published a paper that establishes a model of cortical information processing called hierarchical temporal memory that is based on Bayesian network of Markov chains. They further map this mathematical model to the existing knowledge about the architecture of cortex and show how neurons could recognize patterns by hierarchical Bayesian inference.

Electrophysiology

A number of recent electrophysiological studies focus on the representation of probabilities in the nervous system. Examples are the work of Shadlen and Schultz.

Predictive Coding

Predictive coding is a neurobiologically plausible scheme for inferring the causes of sensory input based on minimizing prediction error. These schemes are related formally to Kalman filtering and other Bayesian update schemes.

Free Energy

During the 1990s some researchers such as Geoffrey Hinton and Karl Friston began examining the concept of free energy as a calculably tractable measure of the discrepancy between actual features of the world and representations of those features captured by neural network models. A synthesis has been attempted recently by Karl Friston, in which the Bayesian brain emerges from a general principle of free energy minimisation. In this framework, both action and perception are seen as a consequence of suppressing free-energy, leading to perceptual and active inference and a more embodied (enactive) view of the Bayesian brain. Using variational Bayesian methods, it can be shown how internal models of the world are updated by sensory information to minimize free energy or the discrepancy between sensory input and predictions of that input. This can be cast (in neurobiologically plausible terms) as predictive coding or, more generally, Bayesian filtering.

According to Friston:

"The free-energy considered here represents a bound on the surprise inherent in any exchange with the environment, under expectations encoded by its state or configuration. A system can minimise free energy by changing its configuration to change the way it samples the environment, or to change its expectations. These changes correspond to action and perception, respectively, and lead to an adaptive exchange with the environment that is characteristic of biological systems. This treatment implies that the system's state and structure encode an implicit and probabilistic model of the environment."

This area of research was summarized in terms understandable by the layperson in a 2008 article in New Scientist that offered a unifying theory of brain function. Friston makes the following claims about the explanatory power of the theory:

"This model of brain function can explain a wide range of anatomical and physiological aspects of brain systems; for example, the hierarchical deployment of cortical areas, recurrent architectures using forward and backward connections and functional asymmetries in these connections. In terms of synaptic physiology, it predicts associative plasticity and, for dynamic models, spike-timing-dependent plasticity. In terms of electrophysiology it accounts for classical and extra-classical receptive field effects and long-latency or endogenous components of evoked cortical responses. It predicts the attenuation of responses encoding prediction error with perceptual learning and explains many phenomena like repetition suppression, mismatch negativity and the P300 in electroencephalography. In psychophysical terms, it accounts for the behavioural correlates of these physiological phenomena, e.g., priming, and global precedence."

"It is fairly easy to show that both perceptual inference and learning rest on a minimisation of free energy or suppression of prediction error."

Mind Uploading

Whole brain emulation (WBE) or mind uploading (sometimes called "mind copying" or "mind transfer") is the hypothetical process of scanning mental state (including long-term memory and "self") of a particular brain substrate and copying it to a computational device, such as a digital, analog, quantum-based or software-based artificial neural network. The computational device could then run a simulation model of the brain's information processing, such that it responds in essentially the same way as the original brain (i.e., indistinguishable from the brain for all relevant purposes) and experiences having a conscious mind.

Mind uploading may potentially be accomplished by either of two methods: Copy-and-Transfer or gradual replacement of neurons. In the case of the former method, mind uploading would be achieved by scanning and mapping the salient features of a biological brain, and then by copying, transferring, and storing that information state into a computer system or another computational device. The simulated mind could be within a virtual reality or simulated world, supported by an anatomic 3D body simulation model. Alternatively, the simulated mind could reside in a computer that is inside (or connected to) a (not necessarily humanoid) robot or a biological body.

Among some futurists and within the transhumanist movement, mind uploading is treated as an important proposed life extension technology. Some believe mind uploading is our current best option for preserving who we are as opposed to cryonics.

Another aim of mind uploading is to provide a permanent backup to our "mind-file", and a means for functional copies of human minds to survive a global disaster or inter-stellar space travels. Whole brain emulation is discussed by some futurists as a "logical endpoint" of the topical computational neuroscience and neuroinformatics fields, both about brain simulation for medical research purposes. It is discussed in artificial intel-ligence research publications as an approach to strong AI. Computer-based intelligence such as an upload could think much faster than a biological human even if it were no more intelligent. A large-scale society of uploads might, according to futurists, give rise to a technological singularity, meaning a sudden time constant decrease in the expo-nential development of technology. Mind uploading is a central conceptual feature of numerous science fiction novels and films.

Substantial mainstream research in related areas is being conducted in animal brain mapping and simulation, development of faster super computers, virtual reality, brain–computer interfaces, connectomics and information extraction from dynamically func-tioning brains. According to supporters, many of the tools and ideas needed to achieve mind uploading already exist or are currently under active development; however, they will admit that others are, as yet, very speculative, but still in the realm of engineering possibility. Neuroscientist Randal Koene has formed a nonprofit organization called Carbon Copies to promote mind uploading research.

Overview

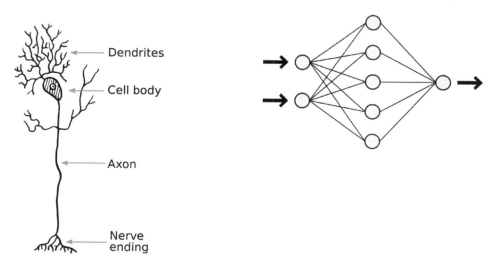

Neuron anatomical model   Simple artificial neural network

The human brain contains about 86 billion nerve cells called neurons, each individually linked to other neurons by way of connectors called axons and dendrites. Signals at the junctures (synapses) of these connections are transmitted by the release and detection of chemicals known as neurotransmitters. The established neuroscientific consensus is that the human mind is largely an emergent property of the information processing of this neural network.

Importantly, neuroscientists have stated that important functions performed by the mind, such as learning, memory, and consciousness, are due to purely physical and electrochemical processes in the brain and are governed by applicable laws. For example, Christof Koch and Giulio Tononi wrote in IEEE Spectrum:

"Consciousness is part of the natural world. It depends, we believe, only on mathematics and logic and on the imperfectly known laws of physics, chemistry, and biology; it does not arise from some magical or otherworldly quality."

The concept of mind uploading is based on this mechanistic view of the mind, and denies the vitalist view of human life and consciousness.

Eminent computer scientists and neuroscientists have predicted that specially programmed computers will be capable of thought and even attain consciousness, including Koch and Tononi, Douglas Hofstadter, Jeff Hawkins, Marvin Minsky, Randal A. Koene, and Rodolfo Llinas.

Such an artificial intelligence capability might provide a computational substrate necessary for uploading.

However, even though uploading is dependent upon such a general capability, it is conceptually distinct from general forms of AI in that it results from dynamic reanimation of information derived from a specific human mind so that the mind retains a sense of historical identity (other forms are possible but would compromise or eliminate the life-extension feature generally associated with uploading). The transferred and reanimated information would become a form of artificial intelligence, sometimes called an infomorph or *"noömorph."*

Many theorists have presented models of the brain and have established a range of estimates of the amount of computing power needed for partial and complete simulations. Using these models, some have estimated that uploading may become possible within decades if trends such as Moore's Law continue.

Theoretical Benefits and Applications

"Immortality" or Backup

In theory, if the information and processes of the mind can be disassociated from the biological body, they are no longer tied to the individual limits and lifespan of that body. Furthermore, information within a brain could be partly or wholly copied or transferred to one or more other substrates (including digital storage or another brain), thereby - from a purely mechanistic perspective - reducing or eliminating "mortality risk" of such information. This general proposal appears to have been first made in the biomedical literature in 1971 by biogerontologist George M. Martin of the University of Washington.

Speedup

If Moore's law holds for several more decades, a supercomputer might be able to simulate a human brain at the neural level at a faster perceived speed than a biological brain. By that time, transistors will have reached sub-atomic size, as current experimental transistors are 10 nm across. However, even if simulation at such speeds should be possible, the exact date this would be achieved is difficult to estimate due to limited understanding of the required accuracy, and computational speed is not the only requirement for making full human brain simulation possible. Several contradictory predictions have been made about when a whole human brain can be emulated, for example 2029 has been suggested by Ray Kurzweil; some of the predicted dates have already passed.

Given that the electrochemical signals that brains use to achieve thought travel at about 150 meters per second, while the electronic signals in computers are sent at 2/3 the speed of light (three hundred million meters per second), this means that a massively parallel electronic counterpart of a human biological brain in theory might be able to think thousands to millions of times faster than our naturally evolved systems. Also, neurons can generate a maximum of about 200 to 1000 action potentials or "spikes" per second, whereas the clock speed of microprocessors has reached 5.5 GHz in 2013, which is about five million times faster.

However, the human brain contains roughly eighty-six billion neurons with eighty-six trillion synapses connecting them. Replicating each of these as separate electronic components using microchip-based semiconductor technology would require a computer enormously large in comparison with today's super-computers. In a less futuristic implementation, time-sharing would allow several neurons to be emulated sequentially by the same computational unit. Thus the size of the computer would be restricted, but the speedup would be lower. Assuming that cortical minicolumns organized into hypercolumns are the computational units, mammal brains can be emulated by today's supercomputers, but with slower speed than in a biological brain.

One obvious use of this technology is the possibility to speed up the development of even faster brains.

Uploaded Astronaut

An "uploaded astronaut" is the application of mind uploading to human spaceflight. An uploaded astronaut would consist of a human mental content transferred or copied to a space humanoid robot or a spacecraft's data storage device. This would eliminate the harms caused by a zero gravity environment, the vacuum of space and cosmic radiation to the human body since both a humanoid robot and a spacecraft can be more resistant than a biological entity in such conditions, permitting longer and farther voyages through outer space than manned spaceflight. Furthermore, an uploaded astronaut may not require a large spacecraft so spacecrafts at the scale of the StarChip might suffice. Alien uploaded astronauts are conceivable as well. In science fiction, Charlie Stross' *Accelerando* features a can-sized starship that visits a nearby star system with an "e-crew" of 63 uploaded astronauts.

Relevant Technologies and Techniques

The focus of mind uploading, in the case of copy-and-transfer, is on data acquisition, rather than data maintenance of the brain. A set of approaches known as loosely coupled off-loading (LCOL) may be used in the attempt to characterize and copy the mental contents of a brain. The LCOL approach may take advantage of self-reports, life-logs and video recordings that can be analyzed by artificial intelligence. A bottom-up approach may focus on the specific resolution and morphology of neurons, the spike times of neurons, the times at which neurons produce action potential responses.

Computational Complexity

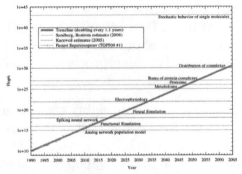

Estimates of how much processing power is needed to emulate a human brain at various levels (from Ray Kurzweil and the chart to the left), along with the fastest supercomputer from TOP500 mapped by year. Note the logarithmic scale and exponential trendline, which assumes the computational capacity doubles every 1.1 years. Kurzweil believes that mind uploading will be possible at neural simulation, while the Sandberg, Bostrom report is less certain about where consciousness arises.

Advocates of mind uploading point to Moore's law to support the notion that the necessary computing power is expected to become available within a few decades. However, the actual computational requirements for running an uploaded human mind are very difficult to quantify, potentially rendering such an argument specious.

Regardless of the techniques used to capture or recreate the function of a human mind, the processing demands are likely to be immense, due to the large number of neurons in the human brain along with the considerable complexity of each neuron.

In 2004, Henry Markram, lead researcher of the "Blue Brain Project", stated that "it is not [their] goal to build an intelligent neural network", based solely on the computational demands such a project would have.

It will be very difficult because, in the brain, every molecule is a powerful computer and we would need to simulate the structure and function of trillions upon trillions of molecules as well as all the rules that govern how they interact. You would literally need computers that are trillions of times bigger and faster than anything existing today.

Five years later, after successful simulation of part of a rat brain, the same scientist was much more bold and optimistic. In 2009, when he was director of the Blue Brain Project, he claimed that

A detailed, functional artificial human brain can be built within the next 10 years

Required computational capacity strongly depend on the chosen level of simulation model scale:

Level	CPU demand (FLOPS)	Memory demand (Tb)	$1 million super-computer (Earliest year of making)
Analog network population model	10^{15}	10^2	2008
Spiking neural network	10^{18}	10^4	2019
Electrophysiology	10^{22}	10^4	2033
Metabolome	10^{25}	10^6	2044
Proteome	10^{26}	10^7	2048
States of protein complexes	10^{27}	10^8	2052
Distribution of complexes	10^{30}	10^9	2063
Stochastic behavior of single molecules	10^{43}	10^{14}	2111
Estimates from *Sandberg, Bostrom, 2008*			

Simulation Model Scale

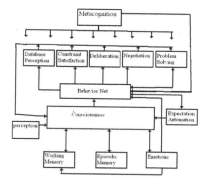

Franklin IDA Architecture

A high-level cognitive AI model of the brain architecture is not required for brain emulation

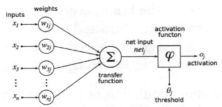

Simple neuron model: Black-box dynamic non-linear signal processing system

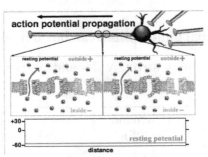

Metabolism model: The movement of positively charged ions through the ion channels controls the membrane electrical action potential in an axon.

Since the function of the human mind, and how it might arise from the working of the brain's neural network, are poorly understood issues, mind uploading relies on the idea of neural network emulation. Rather than having to understand the high-level psychological processes and large-scale structures of the brain, and model them using classical artificial intelligence methods and cognitive psychology models, the low-level structure of the underlying neural network is captured, mapped and emulated with a computer system. In computer science terminology, rather than analyzing and reverse engineering the behavior of the algorithms and data structures that resides in the brain, a blueprint of its source code is translated to another programming language. The human mind and the personal identity then, theoretically, is generated by the emulated neural network in an identical fashion to it being generated by the biological neural network.

On the other hand, a molecule-scale simulation of the brain is not expected to be required, provided that the functioning of the neurons is not affected by quantum mechanical processes. The neural network emulation approach only requires that the functioning and interaction of neurons and synapses are understood. It is expected that it is sufficient with a black-box signal processing model of how the neurons respond to nerve impulses (electrical as well as chemical synaptic transmission).

A sufficiently complex and accurate model of the neurons is required. A traditional artificial neural network model, for example multi-layer perceptron network model, is not considered as sufficient. A dynamic spiking neural network model is required, which reflects that the neuron fires only when a membrane potential reaches a certain level. It is likely that the model must include delays, non-linear functions and differential equations describing the relation between electrophysical parameters such as electrical currents, voltages, membrane states (ion channel states) and neuromodulators.

Since learning and long-term memory are believed to result from strengthening or weakening the synapses via a mechanism known as synaptic plasticity or synaptic adaptation, the model should include this mechanism. The response of sensory receptors to various stimuli must also be modelled.

Furthermore, the model may have to include metabolism, i.e. how the neurons are affected by hormones and other chemical substances that may cross the blood–brain barrier. It is considered likely that the model must include currently unknown neuromodulators, neurotransmitters and ion channels. It is considered unlikely that the simulation model has to include protein interaction, which would make it computationally complex.

A digital computer simulation model of an analog system such as the brain is an approximation that introduces random quantization errors and distortion. However, the biological neurons also suffer from randomness and limited precision, for example due to background noise. The errors of the discrete model can be made smaller than the randomness of the biological brain by choosing a sufficiently high variable resolution and sample rate, and sufficiently accurate models of non-linearities. The computational power and computer memory must however be sufficient to run such large simulations, preferably in real time.

Scanning and Mapping Scale of an Individual

When modelling and simulating the brain of a specific individual, a brain map or connectivity database showing the connections between the neurons must be extracted from an anatomic model of the brain. For whole brain simulation, this network map should show the connectivity of the whole nervous system, including the spinal cord, sensory receptors, and muscle cells. Destructive scanning of a small sample of tissue from a mouse brain including synaptic details is possible as of 2010.

However, if short-term memory and working memory include prolonged or repeated firing of neurons, as well as intra-neural dynamic processes, the electrical and chemical signal state of the synapses and neurons may be hard to extract. The uploaded mind may then perceive a memory loss of the events and mental processes immediately before the time of brain scanning.

A full brain map has been estimated to occupy less than 2×10^{16} bytes (20,000 TB) and would store the addresses of the connected neurons, the synapse type and the synapse "weight" for each of the brains' 10^{15} synapses. However, the biological complexities of true brain function (e.g. the epigenetic states of neurons, protein components with multiple functional states, etc.) may preclude an accurate prediction of the volume of binary data required to faithfully represent a functioning human mind.

Serial Sectioning

A possible method for mind uploading is serial sectioning, in which the brain tissue and perhaps other parts of the nervous system are frozen and then scanned and analyzed layer by layer, which for frozen samples at nano-scale requires a cryo-ultramicrotome, thus capturing the structure of the neurons and their interconnections. The exposed surface of frozen nerve tissue would be scanned and recorded, and then the surface layer of tissue removed. While this would be a very slow and labor-intensive process, research is currently underway to automate the collection and microscopy of serial sections. The scans would then be analyzed, and a model of the neural net recreated in the system that the mind was being uploaded into.

There are uncertainties with this approach using current microscopy techniques. If it is possible to replicate neuron function from its visible structure alone, then the resolution afforded by a scanning electron microscope would suffice for such a technique. However, as the function of brain tissue is partially determined by molecular events (particularly at synapses, but also at other places on the neuron's cell membrane), this may not suffice for capturing and simulating neuron functions. It may be possible to extend the techniques of serial sectioning and to capture the internal molecular makeup of neurons, through the use of sophisticated immunohistochemistry staining methods that could then be read via confocal laser scanning microscopy. However, as the physiological genesis of 'mind' is not currently known, this method may not be able to access all of the necessary biochemical information to recreate a human brain with sufficient fidelity.

Brain Imaging

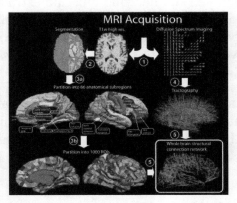

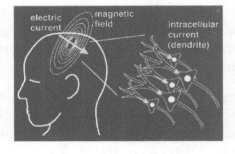

Process from MRI acquisition to whole brain   Magnetoencephalography
            structural network

It may also be possible to create functional 3D maps of the brain activity, using advanced neuroimaging technology, such as functional MRI (fMRI, for mapping change in blood flow), Magnetoencephalography (MEG, for mapping of electrical currents), or combinations of multiple methods, to build a detailed three-dimensional model of the brain using non-invasive and non-destructive methods. Today, fMRI is often combined with MEG for creating functional maps of human cortex during more complex cognitive tasks, as the methods complement each other. Even though current imaging technology lacks the spatial resolution needed to gather the information needed for such a scan, important recent and future developments are predicted to substantially improve both spatial and temporal resolutions of existing technologies.

Brain Simulation

There is ongoing work in the field of brain simulation, including partial and whole simulations of some animals. For example, the *C. elegans* roundworm, Drosophila fruit fly, mouse have all been simulated to various degrees.

The Blue Brain Project by the Brain and Mind Institute of the *École Polytechnique Fédérale de Lausanne,* Switzerland is an attempt to create a synthetic brain by reverse-engineering mammalian brain circuitry.

Issues

Philosophical Issues

Underlying the concept of "mind uploading" (more accurately "mind transferring") is the broad philosophy that consciousness lies within the brain's information processing and is in essence an emergent feature that arises from large neural network high-level patterns of organization, and that the same patterns of organization can be realized in other processing devices. Mind uploading also relies on the idea that the human mind (the "self" and the long-term memory), just like non-human minds, is represented by the current neural network paths and the weights of the brain synapses rather than by a dualistic and mystic soul and spirit. The mind or "soul" can be defined as the information state of the brain, and is immaterial only in the same sense as the information content of a data file or the state of a computer software currently residing in the work-space memory of the computer. Data specifying the information state of the neural network can be captured and copied as a "computer file" from the brain and re-implemented into a different physical form. This is not to deny that minds are richly adapted to their substrates. An analogy to the idea of mind uploading is to copy the temporary information state (the variable values) of a computer program from the computer memory to another computer and continue its execution. The other computer may perhaps have different hardware architecture but emulates the hardware of the first computer.

These issues have a long history. In 1775 Thomas Reid wrote:

I would be glad to know... whether when my brain has lost its original structure, and when some hundred years after the same materials are fabricated so curiously as to become an intelligent being, whether, I say that being will be me; or, if, two or three such beings should be formed out of my brain; whether they will all be me, and consequently one and the same intelligent being.

A considerable portion of transhumanists and singularitarians place great hope into the belief that they may become immortal, by creating one or many non-biological functional copies of their brains, thereby leaving their "biological shell". However, the philosopher and transhumanist Susan Schneider claims that at best, uploading would create a copy of the original persons mind. Susan Schneider agrees that consciousness has a computational basis, but this doesn't mean we can upload and survive. According to her views, "uploading" would probably result in the death of the original person's brain, while only outside observers can maintain the illusion of the original person still being alive. For it is implausible to think that one's consciousness would leave one's brain and travel to a remote location; ordinary physical objects do not behave this way. Ordinary objects (rocks, tables, etc.) are not simultaneously here, and somewhere else. At best, a copy of the original mind is created. Others have argued against such conclusions. For example, buddhist transhumanist James Hughes has pointed out that this consideration only goes so far: if one believes the self is an illusion, worries about survival are not reasons to avoid uploading, and Keith Wiley has presented an argument wherein all resulting minds of an uploading procedure are granted equal primacy in their claim to the original identity, such that survival of the self is determined retroactively from a strictly subjective position.

Another potential consequence of mind uploading is that the decision to "upload" may then create a mindless symbol manipulator instead of a conscious mind. Are we to assume that an Upload is conscious if it displays behaviors that are highly indicative of consciousness? Are we to assume that an Upload is conscious if it verbally insists that it is conscious? Could there be an absolute upper limit in processing speed above which consciousness cannot be sustained? The mystery of consciousness precludes a definitive answer to this question. Numerous scientists, including Kurzweil, strongly believe that determining whether a separate entity is conscious (with 100% confidence) is fundamentally unknowable, since consciousness is inherently subjective. Regardless, some scientists strongly believe consciousness is the consequence of computational processes which are substrateneutral. On the contrary, numerous scientists believe consciousness may be the result of some form of quantum computation dependent on substrate.

In light of uncertainty on whether to regard uploads as conscious, Sandberg proposes a cautious approach:

Principle of assuming the most (PAM): Assume that any emulated system could have the same mental properties as the original system and treat it correspondingly.

It is argued that if a computational copy of one's mind did exist, it would be impossible for one to recognize it as their own mind. The argument for this stance is the following: for a computational mind to recognize an emulation of itself, it must be capable of deciding whether two Turing Machines (namely, itself and the proposed emulation) are functionally equivalent. This task is uncomputable, and thus there cannot exist a computational procedure in the mind that is capable of recognizing an emulation of itself.

Copying vs. Moving

A philosophical issue with mind uploading is whether the newly generated digital mind is really the "same" sentience, or simply an exact copy with the same memories and personality. This issue is especially obvious when the original remains essentially unchanged by the procedure, thereby resulting in a copy which could potentially have rights separate from the unaltered, obvious original.

Most projected brain scanning technologies, such as serial sectioning of the brain, would necessarily be destructive, and the original brain would not survive the brain scanning procedure. But if it can be kept intact, the computer-based consciousness could be a copy of the still-living biological person. It is in that case implicit that copying a consciousness could be as feasible as literally moving it into one or several copies, since these technologies generally involve simulation of a human brain in a computer of some sort, and digital files such as computer programs can be copied precisely. It is assumed that once the versions are exposed to different sensory inputs, their experiences would begin to diverge, but all their memories up until the moment of the copying would remain the same.

The problem is made even more apparent through the possibility of creating a potentially infinite number of initially identical copies of the original person, which would of course all exist simultaneously as distinct beings with their own emotions and thoughts. The most parsimonious view of this phenomenon is that the two (or more) minds would share memories of their past but from the point of duplication would simply be distinct minds.

Toward the goal of resolving the copy-vs-move debate, some have argued for a third way of conceptualizing the process, which is described by such terms as *split* and *divergence*. The distinguishing feature of this third terminological option is that while *moving* implies that a single instance relocates in space and while *copying* invokes problematic connotations (a *copy* is often denigrated in status relative to its *original*), the notion of a *split* better illustrates that some kinds of entities might become two separate instances, but without the imbalanced associations assigned to originals and copies, and that such equality may apply to minds.

Depending on computational capacity, the simulation's subjective time may be faster or slower than elapsed physical time, resulting in that the simulated mind would perceive that the physical world is running in slow motion or fast motion respectively, while biological persons will see the simulated mind in fast or slow motion respectively.

A brain simulation can be started, paused, backed-up and rerun from a saved backup state at any time. The simulated mind would in the latter case forget everything that has happened after the instant of backup, and perhaps not even be aware that it is repeating itself. An older version of a simulated mind may meet a younger version and share experiences with it.

One proposed route for mind uploading is gradual transfer of functions from an "aging biological brain" into an exocortex.

Ethical and Legal Implications

The process of developing emulation technology raises ethical issues related to animal welfare and artificial consciousness. The neuroscience required to develop brain emulation would require animal experimentation, first on invertebrates and then on small mammals before moving on to humans. Sometimes the animals would just need to be euthanized in order to extract, slice, and scan their brains, but sometimes behavioral and *in vivo* measures would be required, which might cause pain to living animals.

In addition, the resulting animal emulations themselves might suffer, depending on one's views about consciousness. Bancroft argues for the plausibility of consciousness in brain simulations on the basis of the "fading qualia" thought experiment of David Chalmers. He then concludes:

If, as I argue above, a sufficiently detailed computational simulation of the brain is potentially operationally equivalent to an organic brain, it follows that we must consider extending protections against suffering to simulations.

It might help reduce emulation suffering to develop virtual equivalents of anaesthesia, as well as to omit processing related to pain and/or consciousness. However, some experiments might require a fully functioning and suffering animal emulation. Animals might also suffer by accident due to flaws and lack of insight into what parts of their brains are suffering. Questions also arise regarding the moral status of *partial* brain emulations, as well as creating neuromorphic emulations that draw inspiration from biological brains but are built somewhat differently.

Brain emulations could be erased by computer viruses or malware, without need to destroy the underlying hardware. This may make assassination easier than for physical humans. The attacker might take the computing power for its own use.

Many questions arise regarding the legal personhood of emulations. Would they be given the rights of biological humans? If a person makes an emulated copy of himself and then dies, does the emulation inherit his property and official positions? Could the emulation ask to "pull the plug" when its biological version was terminally ill or in a coma? Would it help to treat emulations as adolescents for a few years so that the biological creator would maintain temporary control? Would criminal emulations receive the death penalty, or would they be given forced data modification as a form of "rehabilitation"? Could an upload have marriage and child-care rights?

If simulated minds would come true and if they were assigned rights of their own, it may be difficult to ensure the protection of "digital human rights". For example, social science researchers might be tempted to secretly expose simulated minds, or whole isolated societies of simulated minds, to controlled experiments in which many copies of the same minds are exposed (serially or simultaneously) to different test conditions.

Political and Economic Implications

Emulations could create a number of conditions that might increase risk of war, including inequality, changes of power dynamics, a possible technological arms race to build emulations first, first-strike advantages, strong loyalty and willingness to "die" among emulations, and triggers for racist, xenophobic, and religious prejudice. If emulations run much faster than humans, there might not be enough time for human leaders to make wise decisions or negotiate. It's possible that humans would react violently against growing power of emulations, especially if they depress human wages. Or maybe emulations wouldn't trust each other, and even well intentioned defensive measures might be interpreted as offense.

Emulation Timelines and AI risk

There are very few feasible technologies that humans have refrained from developing. The neuroscience and computer-hardware technologies that may make brain emulation possible are widely desired for other reasons, so cutting off funding doesn't seem to be an option. If we assume that emulation technology will arrive, a question becomes whether we should accelerate or slow its advance.

Arguments for speeding up brain-emulation research:

- If neuroscience is the bottleneck on brain emulation rather than computing power, emulation advances may be more erratic and unpredictable based on when new scientific discoveries happen. Limited computing power would mean the first emulations would run slower and so would be easier to adapt to, and there would be more time for the technology to transition through society.

- Improvements in manufacturing, 3D printing, and nanotechnology may accelerate hardware production, which could increase the "computing overhang" from excess hardware relative to neuroscience.

- If one AI-development group had a lead in emulation technology, it would have more subjective time to win an arms race to build the first superhuman AI. Because it would be less rushed, it would have more freedom to consider AI risks.

Arguments for slowing down brain-emulation research:

- Greater investment in brain emulation and associated cognitive science might enhance the ability of artificial intelligence (AI) researchers to create "neuromorphic" (brain-inspired) algorithms, such as neural networks, reinforcement learning, and hierarchical perception. This could accelerate risks from uncontrolled AI. Participants at a 2011 AI workshop estimated an 85% probability that neuromorphic AI would arrive before brain emulation. This was based on the idea that brain emulation would require understanding some brain components, and it would be easier to tinker with these than to reconstruct the entire brain in its original form. By a very narrow margin, the participants on balance leaned toward the view that accelerating brain emulation would increase expected AI risk.

- Waiting might give society more time to think about the consequences of brain emulation and develop institutions to improve cooperation.

Emulation research would also speed up neuroscience as a whole, which might accelerate medical advances, cognitive enhancement, lie detectors, and capability for psychological manipulation.

Emulations might be easier to control than *de novo* AI because

1. We understand better human abilities, behavioral tendencies, and vulnerabilities, so control measures might be more intuitive and easier to plan for.

2. Emulations could more easily inherit human motivations.

3. Emulations are harder to manipulate than *de novo* AI, because brains are messy and complicated; this could reduce risks of their rapid takeoff. Also, emulations may be bulkier and require more hardware than AI, which would also slow the speed of a transition. Unlike AI, an emulation wouldn't be able to rapidly expand beyond the size of a human brain. Emulations running at digital speeds would have less intelligence differential vis-à-vis AI and so might more easily control AI.

As counterpoint to these considerations, Bostrom notes some downsides:

1. Even if we better understand human behavior, the *evolution* of emulation behavior under self-improvement might be much less predictable than the evolution of safe *de novo* AI under self-improvement.

2. Emulations may not inherit all human motivations. Perhaps they would inherit our darker motivations or would behave abnormally in the unfamiliar environment of cyberspace.

3. Even if there's a slow takeoff toward emulations, there would still be a second transition to *de novo* AI later on. Two intelligence explosions may mean more total risk.

Mind Uploading Advocates

Ray Kurzweil, director of engineering at Google, claims to know and foresee that people will be able to "upload" their entire brains to computers and become "digitally immortal" by 2045. Kurzweil made this claim for many years, e.g. during his speech in

2013 at the Global Futures 2045 International Congress in New York, which claims to subscribe to a similar set of beliefs. Mind uploading is also advocated by a number of researchers in neuroscience and artificial intelligence, such as Marvin Minsky while he was still alive. In 1993, Joe Strout created a small web site called the Mind Uploading Home Page, and began advocating the idea in cryonics circles and elsewhere on the net. That site has not been actively updated in recent years, but it has spawned other sites including MindUploading.org, run by Randal A. Koene, Ph.D., who also moderates a mailing list on the topic. These advocates see mind uploading as a medical procedure which could eventually save countless lives.

Many transhumanists look forward to the development and deployment of mind uploading technology, with transhumanists such as Nick Bostrom predicting that it will become possible within the 21st century due to technological trends such as Moore's Law.

Michio Kaku, in collaboration with Science hosted a documentary, Sci Fi Science: Physics of the Impossible based on his book Physics of the Impossible and in episode 4 on "How to teleport" mentions that mind uploading via techniques such as quantum entanglement and whole brain emulation using an advanced MRI machine may enable people to be transported to vast distances at near light-speed.

The book *Beyond Humanity: CyberEvolution and Future Minds* by Gregory S. Paul & Earl D. Cox, is about the eventual (and, to the authors, almost inevitable) evolution of computers into sentient beings, but also deals with human mind transfer. Richard Doyle's *Wetwares: Experiments in PostVital Living* deals extensively with uploading from the perspective of distributed embodiment, arguing for example that humans are currently part of the "artificial life phenotype." Doyle's vision reverses the polarity on uploading, with artificial life forms such as uploads actively seeking out biological embodiment as part of their reproductive strategy.

Mind Uploading Skeptics

Kenneth D. Miller, a professor of neuroscience at Columbia and a co-director of the Center for Theoretical Neuroscience, raised doubts about the practicality of mind uploading.

Neurocomputational Speech Processing

Neurocomputational speech processing is computer-simulation of speech production and speech perception by referring to the natural neuronal processes of speech production and speech perception, as they occur in the human nervous system (central nervous system and peripheral nervous system). This topic is based on neuroscience and computational neuroscience.

Overview

Neurocomputational models of speech processing are complex. They comprise at least a cognitive part, a motor part and a sensory part.

The cognitive or linguistic part of a neurocomputational model of speech processing comprises the neural activation or generation of a phonemic representation on the side of speech production (e.g. neurocomputational and extended version of the Levelt model developed by Ardi Roelofs: WEAVER++ as well as the neural activation or generation of an intention or meaning on the side of speech perception or speech comprehension.

The motor part of a neurocomputational model of speech processing starts with a phonemic representation of a speech item, activates a motor plan and ends with the articulation of that particular speech item.

The sensory part of a neurocomputational model of speech processing starts with an acoustic signal of a speech item (acoustic speech signal), generates an auditory representation for that signal and activates a phonemic representations for that speech item.

Neurocomputational Speech Processing Topics

Neurocomputational speech processing is speech processing by artificial neural networks. Neural maps, mappings and pathways as described below, are model structures, i.e. important structures within artificial neural networks.

Neural Maps

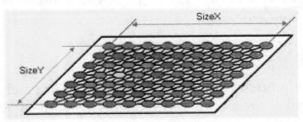

Fig. 2D neuronal map with a local activation pattern. magenta: neuron with highest degree of activation; blue: neurons with no activation

A neural network can be separated in three types of neural maps, also called "layers":

1. input maps (in the case of speech processing: primary auditory map within the auditory cortex, primary somatosensory map within the somatosensory cortex),

2. output maps (primary motor map within the primary motor cortex), and

3. higher level cortical maps (also called "hidden layers", see neural networks).

The term "neural map" is favoured here over the term "neural layer", because a cortial neural map should be modeled as a 2D-map of interconnected neurons (e.g. like a self-organizing map; see also Fig. 1). Thus, each "model neuron" or "artificial neuron" within this 2D-map is physiologically represented by a cortical column since the cerebral cortex anatomically exhibits a layered structure.

Neural Representations (Neural States)

A neural representation within an artificial neural network is a temporarily activated (neural) state within a specific neural map. Each neural state is represented by a specific neural activation pattern. This activation pattern changes during speech processing (e.g. from syllable to syllable).

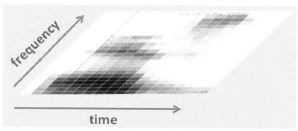

Fig. 2D neuronal map with a distributed activation pattern. Example: "neural spectrogram" (This auditory neural representation is speculative; ACT model, below)

In the ACT model, it is assumed that an auditory state can be represented by a "neural spectrogram" (see Fig.) within an auditory state map. This auditory state map is assumed to be located in the auditory association cortex.

A somatosensory state can be divided in a tactile and proprioceptive state and can be represented by a specific neural activation pattern within the somatosensory state map. This state map is assumed to be located in the somatosensory association cortex (see cerebral cortex, somatosensory system, somatosensory cortex).

A motor plan state can be assumed for representing a motor plan, i.e. the planning of speech articulation for a specific syllable or for a longer speech item (e.g. word, short phrase). This state map is assumed to be located in the premotor cortex, while the instantaneous (or lower level) activation of each speech articulator occurs within the primary motor cortex.

The neural representations occurring in the sensory and motor maps (as introduced above) are distributed representations (Hinton et al. 1968): Each neuron within the sensory or motor map is more or less activated, leading to a specific activation pattern.

The neural representation for speech units occurring in the speech sound map is a punctual or local representation. Each speech item or speech unit is represented here by a specific neuron.

Neural Mappings (Synaptic Projections)

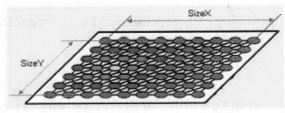

Fig. Neural mapping between phonetic map (local activation pattern for a specific phonetic state), motor plan state map (distributed activation pattern) and auditory state map (distributed activation pattern) as part of the ACT model. Only neural connections with the winner neuron within the phonetic map are shown

A neural mapping connects two cortical neural maps. Neural mappings (in contrast to neural pathways) store training information by adjusting their neural link weights. Neural mappings are capable of generating or activating a distributed representation of a sensory or motor state within a sensory or motor map from a punctual or local activation within the other map (for example the synaptic projection from speech sound map to motor map, to auditory target region map, or to somatosensory target region map in the DIVA model, explained below; or see for example the neural mapping from phonetic map to auditory state map and motor plan state map in the ACT model, explained below and Fig.).

Neural mapping between two neural maps are compact or dense: Each neuron of one neural map is interconnected with (nearly) each neuron of the other neural map (many-to-many-connection). Because of this density criterion for neural mappings, neural maps which are interconnected by a neural mapping are not far apart from each other.

Neural Pathways

In contrast to neural mappings neural pathways can connect neural maps which are far apart (e.g. in different cortical lobes). From the functional or modeling viewpoint, neural pathways mainly forward information without processing this information. A neural pathway in comparison to a neural mapping need much less neural connections. A neural pathway can be modelled by using a one-to-one connection of the neurons of both neural maps.

Example: In the case of two neural maps, each comprising 1,000 model neurons, a neural mapping needs up to 1,000,000 neural connections (many-to-many-connection), while only 1,000 connections are needed in the case of a neural pathway connection.

Furthermore the link weights of the connections within a neural mapping are adjusted during training, while the neural connections in the case of a neural pathway need not to be trained (each connection is maximal exhibitory).

DIVA Model

The leading approach in neurocomputational modeling of speech production is the DIVA model developed by Frank H. Guenther and his group at Boston University. The model accounts for a wide range of phonetic and neuroimaging data but - like each neurocomputational model - remains speculative to some extent.

Structure of the Model

The organization or structure of the DIVA model is shown in Fig.

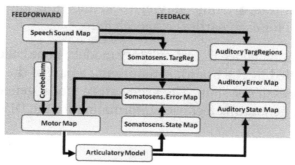

Fig. Organization of the DIVA model; This figure is an adaptation following Guenther et al. 2006

Speech Sound Map: The Phonemic Representation as a Starting Point

The speech sound map - assumed to be located in the inferior and posterior portion of Broca's area (left frontal operculum) - represents (phonologically specified) language-specific speech units (sounds, syllables, words, short phrases). Each speech unit (mainly syllables; e.g. the syllable and word "palm" /pam/, the syllables /pa/, /ta/, /ka/, ...) is represented by a specific model cell within the speech sound map (i.e. punctual neural representations). Each model cell corresponds to a small population of neurons which are located at close range and which fire together.

Feedforward Control: Activating Motor Representations

Each neuron (model cell, artificial neuron) within the speech sound map can be activated and subsequently activates a forward motor command towards the motor map, called articulatory velocity and position map. The activated neural representation on the level of that motor map determines the articulation of a speech unit, i.e. controls all articulators (lips, tongue, velum, glottis) during the time interval for producing that speech unit. Forward control also involves subcortical structures like the cerebellum, not modelled in detail here.

A speech *unit* represents an amount of speech *items* which can be assigned to the same phonemic category. Thus, each speech unit is represented by one specific neuron within the speech sound map, while the realization of a speech unit may exhibit some articulatory and acoustic variability. This phonetic variability is the motivation to define sensory target *regions* in the DIVA model.

Articulatory Model: Generating Somatosensory and Auditory Feedback Information

The activation pattern within the motor map determines the movement pattern of all model articulators (lips, tongue, velum, glottis) for a speech item. In order not to overload the model, no detailed modeling of the neuromuscular system is done. The Maeda articulatory speech synthesizer is used in order to generate articulator movements, which allows the generation of a time-varying vocal tract form and the generation of the acoustic speech signal for each particular speech item.

In terms of artificial intelligence the articulatory model can be called plant (i.e. the system, which is controlled by the brain); it represents a part of the embodiment of the neuronal speech processing system. The articulatory model generates sensory output which is the basis for generating feedback information for the DIVA model.

Feedback Control: Sensory Target Regions, State Maps, and Error Maps

On the one hand the articulatory model generates sensory information, i.e. an auditory state for each speech unit which is neurally represented within the auditory state map (distributed representation), and a somatosensory state for each speech unit which is neurally represented within the somatosensory state map (distributed representation as well). The auditory state map is assumed to be located in the superior temporal cortex while the somatosensory state map is assumed to be located in the inferior parietal cortex.

On the other hand the speech sound map, if activated for a specific speech unit (single neuron activation; punctual activation), activates sensory information by synaptic projections between speech sound map and auditory target region map and between speech sound map and somatosensory target region map. Auditory and somatosensory target regions are assumed to be located in higher-order auditory cortical regions and in higher-order somatosensory cortical regions respectively. These target region sensory activation patterns - which exist for each speech unit - are learned during speech acquisition.

Consequently two types of sensory information are available if a speech unit is activated at the level of the speech sound map: (i) learned sensory target regions (i.e. *intended* sensory state for a speech unit) and (ii) sensory state activation patterns resulting from a possibly imperfect execution (articulation) of a specific speech unit (i.e. *current* sensory state, reflecting the current production and articulation of that particular speech unit). Both types of sensory information is projected to sensory error maps, i.e. to an auditory error map which is assumed to be located in the superior temporal cortex (like the auditory state map) and to a somatosensosry error map which is assumed to be located in the inferior parietal cortex (like the somatosensory state map) (see Fig.).

If the current sensory state deviates from the intended sensory state, both error maps are generating feedback commands which are projected towards the motor map and which are capable to correct the motor activation pattern and subsequently the articulation of a speech unit under production. Thus, in total, the activation pattern of the motor map is not only influenced by a specific feedforward command learned for a speech unit (and generated by the synaptic projection from the speech sound map) but also by a feedback command generated at the level of the sensory error maps (see Fig.).

Learning (Modeling Speech Acquisition)

While the *structure* of a neuroscientific model of speech processing (given in Fig. for the DIVA model) is mainly determined by evolutionary processes, the (language-specific) *knowledge* as well as the (language-specific) *speaking skills* are learned and trained during speech acquisition. In the case of the DIVA model it is assumed that the newborn has not available an already structured (language-specific) speech sound map; i.e. no neuron within the speech sound map is related to any speech unit. Rather the organization of the speech sound map as well as the tuning of the projections to the motor map and to the sensory target region maps is learned or trained during speech acquisition. Two important phases of early speech acquisition are modeled in the DIVA approach: Learning by babbling and by imitation.

Babbling

During babbling the synaptic projections between sensory error maps and motor map are tuned. This training is done by generating an amount of semi-random feedforward commands, i.e. the DIVA model "babbles". Each of these babbling commands leads to the production of an "articulatory item", also labeled as "pre-linguistic (i.e. non language-specific) speech item" (i.e. the articulatory model generates an articulatory movement pattern on the basis of the babbling motor command). Subsequently an acoustic signal is generated.

On the basis of the articulatory and acoustic signal, a specific auditory and somatosensory state pattern is activated at the level of the sensory state maps (see Fig.) for each (pre-linguistic) speech item. At this point the DIVA model has available the sensory and associated motor activation pattern for different speech items, which enables the model to tune the synaptic projections between sensory error maps and motor map. Thus, during babbling the DIVA model learns feedback commands (i.e. how to produce a proper (feedback) motor command for a specific sensory input).

Imitation

During imitation the DIVA model organizes its speech sound map and tunes the synaptic projections between speech sound map and motor map - i.e. tuning of forward motor commands - as well as the synaptic projections between speech sound map and sensory target regions (see Fig.). Imitation training is done by exposing the model to an amount of acoustic speech signals representing realizations of language-specific speech units (e.g. isolated speech sounds, syllables, words, short phrases).

The tuning of the synaptic projections between speech sound map and auditory target region map is accomplished by assigning one neuron of the speech sound map to the phonemic representation of that speech item and by associating it with the auditory representation of that speech item, which is activated at the auditory target region map. Auditory *regions* (i.e. a specification of the auditory vairiability of a speech unit) occur, because one specific speech unit (i.e. one specific phonemic representation) can be realized by several (slightly) different acoustic (auditory) realizations (for the difference between speech *item* and speech *unit* see above: feedforward control) .

The tuning of the synaptic projections between speech sound map and motor map (i.e. tunig of forward motor commands) is accomplished with the aid of feedback commands, since the projections between sensory error maps and motor map were already tuned during babbling training (see above). Thus the DIVA model tries to "imitate" an auditory speech item by attempting to find a proper feedforward motor command. Subsequently the model compares the resulting sensory output (*current* sensory state following the articulation of that attempt) with the already learned auditory target region (*intended* sensory state) for that speech item. Then the model updates the current feedforward motor command by the current feedback motor command generated from the auditory error map of the auditory feedback system. This process may be repeated several times (several attempts). The DIVA model is capable of producing the speech item with a decreasing auditory difference between curren and intended auditory state from attempt to attempt.

During imitation the DIVA model is also capable of tuning the synaptic projections from speech sound map to somatosensory target region map, since each new imitation attempt produces a new articulation of the speech item and thus produces a somatosensory state pattern which is associated with the phonemic representation of that speech item.

Perturbation Experiments

Real-time Perturbation of F1: the Influence of Auditory Feedback

While auditory feedback is most important during speech acquisition, it may be activated less if the model has learned a proper feedforward motor command for each speech unit. But it has been shown that auditory feedback needs to be strongly coactivated in the case of auditory perturbation (e.g. shifting a formant frequency, Tourville et al. 2005). This is comparable to the strong influence of visual feedback on reaching movements during visual perturbation (e.g. shifting the location of objects by viewing through a prism).

Unexpected Blocking of the Jaw: The Influence of Somatosensory Feedback

In a comparable way to auditory feedback, also somatosensory feedback can be strongly coactivated during speech production, e.g. in the case of unexpected blocking of the jaw (Tourville et al. 2005).

ACT Model

A further approach in neurocomputational modeling of speech processing is the ACT model developed by Bernd J. Kröger and his group at RWTH Aachen University, Germany (Kröger et al. 2014, Kröger et al. 2009, Kröger et al. 2011). The ACT model is in accord with the DIVA model in large parts. The ACT model focuses on the "action repository" (i.e. repository for sensorimotor speaking skills, comparable to the mental syllablary, Levelt and Wheeldon 1994), which is not spelled out in detail in the DIVA model. Moreover the ACT model explicitly introduces a level of motor plans, i.e. a high-level motor description for the production of speech items. The ACT model - like any neurocomputational model - remains speculative to some extent.

Structure

The organization or structure of the ACT model is given in Fig.

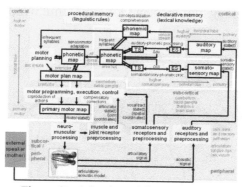

Fig. 5: Organization of the ACT model

For speech production, the ACT model starts with the activation of a phonemic representation of a speech item (phonemic map). In the case of a *frequent syllable*, a co-activation occurs at the level of the phonetic map, leading to a further co-activation of the intended sensory state at the level of the sensory state maps and to a co-activation of a motor plan state at the level of the motor plan map. In the case of an *infrequent syllable*, an attempt for a motor plan is generated by the motor planning module for that speech item by activating motor plans for phonetic similar speech items via the phonetic map. The motor plan or vocal tract action score comprises temporally overlapping vocal tract actions, which are programmed and subsequently executed by the motor programming, execution, and control module. This module gets real-time somatosensory feedback information for controlling the correct execution of the (intended) motor plan. Motor programing leads to activation pattern at the level lof the primary motor map and subsequently activates neuromuscular processing. Motoneuron activation patterns generate muscle forces and subsequently movement patterns of all model articulators (lips, tongue, velum, glottis). The Birkholz 3D articulatory synthesizer is used in order to generate the acoustic speech signal.

Articulatory and acoustic feedback signals are used for generating somatosensory and auditory feedback information via the sensory preprocessing modules, which is forwarded towards the auditory and somatosensory map. At the level of the sensory-phonetic processing modules, auditory and somatosensory information is stored in short-term memory and the external sensory signal (ES, Fig., which are activated via the sensory feedback loop) can be compared with the already trained sensory signals (TS, Fig., which are activated via the phonetic map). Auditory and somatosensory error signals can be generated if external and intended (trained) sensory signals are noticeably different (cf. DIVA model).

The light green area in Fig. indicates those neural maps and processing modules, which process a syllable as a whole unit (specific processing time window around 100 ms and more). This processing comprises the phonetic map and the directly connected sensory state maps within the sensory-phonetic processing modules and the directly connected motor plan state map, while the primary motor map as well as the (primary) auditory and (primary) somatosensory map process smaller time windows (around 10 ms in the ACT model).

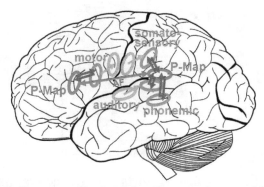

Fig. Hypothetical location of brain regions for neural maps of the ACT model

The hypothetical cortical location of neural maps within the ACT model is shown in Fig. The hypothetical locations of primary motor and primary sensory maps are given in magenta, the hypothetical locations of motor plan state map and sensory state maps (within sensory-phonetic processing module, comparable to the error maps in DIVA) are given in orange, and the hypothetical locations for the mirrored phonetic map is given in red. Double arrows indicate neuronal mappings. Neural mappings connect neural maps, which are not far apart from each other. The two mirrored locations of the phonetic map are connected via a neural pathway, leading to a (simple) one-to-one mirroring of the current activation pattern for both realizations of the phonetic map. This neural pathway between the two locations of the phonetic map is assumed to be a part of the fasciculus arcuatus.

For speech perception, the model starts with an external acoustic signal (e.g. produced by an external speaker). This signal is preprocessed, passes the auditory map, and leads to an activation pattern for each syllable or word on the level of the auditory-phonetic processing module. The ventral path of speech perception would directly activate a lexical item, but is not implemented in ACT. Rather, in ACT the activation of a phonemic state occurs via the phonemic map and thus may lead to a coactivation of motor representations for that speech item (i.e. dorsal pathway of speech perception; ibid.).

Action Repository

Fig. Visualization of synaptic link weights for a section of the phonetic map, trained for the 200 most frequent syllables of Standard German. Each box represents a neuron within the self-organizing phonetic map. Each of the three link weight representations refers to the same section within the phonetic map and thus refers to the same 10×10 neurons

The phonetic map together with the motor plan state map, sensory state maps (occurring within the sensory-phonetic processing modules), and phonemic (state) map form the action repository. The phonetic map is implemented in ACT as a self-organizing neural map and different speech items are represented by different neurons within this map (punctual or local representation, see above: neural representations). The phonetic map exhibits three major characteristics:

- More than one phonetic realization may occur within the phonetic map for one phonemic state (phonemic link weights in Fig. e.g. the syllable /de:m/ is represented by three neurons within the phonetic map)

- Phonetotopy: The phonetic map exhibits an ordering of speech items with respect to different phonetic features (phonemic link weights in Fig. Three examples: (i) the syllables /p@/, /t@/, and /k@/ occur in an upward ordering at the left side within the phonetic map; (ii) syllable-initial plosives occur in the upper left part of the phonetic map while syllable initial fricatives occur in the lower right half; (iii) CV syllables and CVC syllables as well occur in different areas of the phonetic map.).

- The phonetic map is hypermodal or multimodal: The activation of a phonetic item at the level of the phonetic map coactivates (i) a phonemic state (phonemic link weights in Fig.), (ii) a motor plan state (motor plan link weights in Fig.), (iii) an auditory state (auditory link weights in Fig.), and (iv) a somatosensory state (not shown in Fig. 7). All these states are learned or trained during speech acquisition by tuning the synaptic link weights between each neuron within the phonetic map, representing a particular phonetic state and all neurons within the associated motor plan and sensory state maps.

The phonetic map implements the action-perception-link within the ACT model (The dual neural representation of the phonetic map in the frontal lobe and at the intersection of temporal lobe and parietal lobe).

Motor Plans

A motor plan is a high level motor description for the production and articulation of a speech items. In our neurocomputational model ACT a motor plan is quantified as a vocal tract action score. Vocal tract action scores quantitatively determine the number of vocal tract actions (also called articulatory gestures), which need to be activated in order to produce a speech item, their degree of realization and duration, and the temporal organization of all vocal tract actions building up a speech item. The detailed realization of each vocal tract action (articulatory gesture) depends on the temporal organization of all vocal tract actions building up a speech item and especially on their temporal overlap. Thus the detailed realization of each vocal tract action within an speech item is specified below the motor plan level in our neurocomputational model ACT.

Integrating Sensorimotor and Cognitive Aspects: The Coupling of Action Repository and Mental Lexicon

A severe problem of phonetic or sensorimotor models of speech processing (like DIVA or ACT) is that the development of the phonemic map during speech acquisition is not modeled. A possible solution of this problem could be a direct coupling of action repository and mental lexicon without explicitly introducing a phonemic map at the beginning of speech acquisition.

Experiments: Speech Acquisition

A very important issue for all neuroscientific or neurocomputational approaches is to separate structure and knowledge. While the structure of the model (i.e. of the human neuronal network, which is needed for processing speech) is mainly determined by evolutionary processes, the knowledge is gathered mainly during speech acquisition by processes of learning. Different learning experiments were carried out with the model ACT in order to learn (i) a five-vowel system /i, e, a, o, u/, (ii) a small consonant system (voiced plosives /b, d, g/ in combination with all five vowels acquired earlier as CV syllables (ibid.), (iii) a small model language comprising the five-vowel system, voiced and unvoiced plosives /b, d, g, p, t, k/, nasals /m, n/ and the lateral /l/ and three syllable types (V, CV, and CCV) and (iv) the 200 most frequent syllables of Standard German for a 6 year old child. In all cases, an ordering of phonetic items with respect to different phonetic features can be observed.

Experiments: Speech Perception

Despite the fact that the ACT model in its earlier versions was designed as a pure speech production model (including speech acquisition), the model is capable of exhibiting important basic phenomena of speech perception, i.e. categorical perception and the McGurk effect. In the case of categorical perception, the model is able to exhibit that categorical perception is stronger in the case of plosives than in the case of vowels. Furthermore the model ACT was able to exhibit the McGurk effect, if a specific mechanism of inhibition of neurons of the level of the phonetic map was implemented.

Brain-Reading

Brain-reading uses the responses of multiple voxels in the brain evoked by stimulus then detected by fMRI in order to decode the original stimulus. Brain reading studies differ in the type of decoding (i.e. classification, identification and reconstruction) employed, the target (i.e. decoding visual patterns, auditory patterns, cognitive states), and the decoding algorithms (linear classification, nonlinear classification, direct reconstruction, Bayesian reconstruction, etc.) employed.

Classification

In classification, a pattern of activity across multiple voxels is used to determine the particular class from which the stimulus was drawn. Many studies have classified visual stimuli, but this approach has also been used to classify cognitive states.

Reconstruction

In reconstruction brain reading the aim is to create a literal picture of the image that was presented. Early studies used voxels from early visual cortex areas (V1, V2, and V3) to reconstruct geometric stimuli made up of flickering checkerboard patterns.

Natural Images

More recent studies used voxels from early and anterior visual cortex areas forward of them (visual areas V3A, V3B, V4, and the lateral occipital) together with Bayesian inference techniques to reconstruct complex natural images. This brain reading approach uses three components: A structural encoding model that characterizes responses in early visual areas; a semantic encoding model that characterizes responses in anterior visual areas; and a Bayseian prior that describes the distribution of structural and semantic scene statistics.

Experimentally the procedure is for subjects to view 1750 black and white natural images that are correlated with voxel activation in their brains. Then subjects viewed another 120 novel target images, and information from the earlier scans is used reconstruct them. Natural images used include pictures of a seaside cafe and harbor, performers on a stage, and dense foliage.

Other Types

It is possible to track which of two forms of rivalrous binocular illusions a person was subjectively experiencing from fMRI signals. The category of event which a person freely recalls can be identified from fMRI before they say what they remembered. Statistical analysis of EEG brainwaves has been claimed to allow the recognition of phonemes, and at a 60% to 75% level color and visual shape words. It has also been shown that brain-reading can be achieved in a complex virtual environment.

Accuracy

Brain-reading accuracy is increasing steadily as the quality of the data and the complexity of the decoding algorithms improve. In one recent experiment it was possible to identify which single image was being seen from a set of 120. In another it was possible to correctly identify 90% of the time which of two categories the stimulus came and the specific semantic category (out of 23) of the target image 40% of the time.

Limitations

It has been noted that so far brain reading is limited. "In practice, exact reconstructions are impossible to achieve by any reconstruction algorithm on the basis of brain activity signals acquired by fMRI. This is because all reconstructions will inevitably be limited by inaccuracies in the encoding models and noise in the measured signals. Our results demonstrate that the natural image prior is a powerful (if unconventional) tool for mitigating the effects of these fundamental limitations. A natural image prior with only six million images is sufficient to produce reconstructions that are structurally and semantically similar to a target image."

Applications

Brain reading has been suggested as an alternative to polygraph machines as a form of lie detection. One neuroimaging method that has been proposed as a lie detector is EEG "brain-fingerprinting", in which event related potentials are supposedly used to determine whether a stimulus is familiar or unfamiliar. The inventor of brain fingerprinting, Lawrence Farwell, has attempted to demonstrate its use in a legal case, Harrington v. State of Iowa, although the state objected on the basis that the probes used by Farwell were too general for familiarity or unfamiliarity with them to prove innocence. Another alternative to polygraph machines is Blood Oxygenated Level Dependent functional MRI technology (BOLD fMRI). This technique involves the interpretation of the local change in the concentration of oxygenated hemoglobin in the brain, although the relationship between this blood flow and neural activity is not yet completely understood.

A number of concerns have been raised about the accuracy and ethical implications of brain reading for this purpose. Laboratory studies have found rates of accuracy of up to 85%; however, there are concerns about what this means for false positive results among non-criminal populations: "If the prevalence of "prevaricators" in the group being examined is low, the test will yield far more false-positive than true-positive results; about one person in five will be incorrectly identified by the test." Ethical problems involved in the use of brain reading as lie detection include misapplications due to adoption of the technology before its reliability and validity can be properly assessed and due to misunderstanding of the technology, and privacy concerns due to unprecedented access to individual's private thoughts. However, it has been noted that the use of polygraph lie detection carries similar concerns about the reliability of the results and violation of privacy.

Brain-reading has also been proposed as a method of improving human-machine interfaces, by the use of EEG to detect relevant brain states of a human. In recent years, there has been a rapid increase in patents for technology involved in reading brain-waves, rising from fewer than 400 from 2009-2012 to 1600 in 2014. These include proposed ways to control video games via brain waves and "neuro-marketing" to determine someone's thoughts about a new product or advertisement.

References

- Naumann, J. Search for Paradise: A Patient's Account of the Artificial Vision Experiment (2012), Xlibris Corporation, ISBN 1-479-7092-04

- Gürkök H., Nijholt A., Poel M. (2012). "Brain-Computer Interface Games: Towards a Framework". ICEC. Lecture Notes in Computer Science. 2012: 373–380. doi:10.1007/978-3-642-33542-6_33. ISBN 978-3-642-33541-9.

- Wiley, Keith (Sep 2014). A Taxonomy and Metaphysics of Mind-Uploading (1st ed.). Humanity+ Press and Alautun Press. ISBN 978-0692279847. Retrieved 16 October 2014.

- Bostrom, Nick (2014). "Ch. 14: The strategic picture". Superintelligence: Paths, Dangers, Strategies. Oxford University Press. ISBN 978-0199678112.

- Prisco, Giulio (12 December 2012). "Uploaded e-crews for interstellar missions". kurzweilai.net. Retrieved 31 July 2015.

- Wiley, Keith (March 20, 2014). "Response to Susan Schneider's "Philosophy of 'Her'"". H+Magazine. Retrieved 7 May 2014.

- Shulman, Carl; Anders Sandberg (2010). Mainzer, Klaus, ed. "Implications of a Software-Limited Singularity" (PDF). ECAP10: VIII European Conference on Computing and Philosophy. Retrieved 17 May 2014.

- Anna Salamon; Luke Muehlhauser (2012). "Singularity Summit 2011 Workshop Report" (PDF). Machine Intelligence Research Institute. Retrieved 28 June 2014.

5

Allied Disciplines of Computational Neuroscience

The research findings of computational neuroscience are utilized in a wide array of allied disciplines like computational anatomy, neuroethology, neural coding, neuroinformatics, neurocybernetics etc. This section explores these fields in-depth and provides a brief background about each. The chapter is a compilation of the various allied branches of computational neurosciences that form an integral part of the broader subject matter.

Machine Learning

Machine learning is a subfield of computer science (more particularly soft computing) that evolved from the study of pattern recognition and computational learning theory in artificial intelligence. In 1959, Arthur Samuel defined machine learning as a "Field of study that gives computers the ability to learn without being explicitly programmed". Machine learning explores the study and construction of algorithms that can learn from and make predictions on data. Such algorithms operate by building a model from example inputs in order to make data-driven predictions or decisions, rather than following strictly static program instructions.

Machine learning is closely related to (and often overlaps with) computational statistics; a discipline which also focuses in prediction-making through the use of computers. It has strong ties to mathematical optimization, which delivers methods, theory and application domains to the field. Machine learning is employed in a range of computing tasks where designing and programming explicit algorithms is unfeasible. Example applications include spam filtering, optical character recognition (OCR), search engines and computer vision. Machine learning is sometimes conflated with data mining, where the latter sub-field focuses more on exploratory data analysis and is known as unsupervised learning.

Within the field of data analytics, machine learning is a method used to devise complex models and algorithms that lend themselves to prediction - in commercial use, this is known as predictive analytics. These analytical models allow researchers, data scientists, engineers, and analysts to "produce reliable, repeatable decisions and results" and uncover "hidden insights" through learning from historical relationships and trends in the data.

Overview

Tom M. Mitchell provided a widely quoted, more formal definition: "A computer program is said to learn from experience E with respect to some class of tasks T and performance measure P if its performance at tasks in T, as measured by P, improves with experience E." This definition is notable for its defining machine learning in fundamentally operational rather than cognitive terms, thus following Alan Turing's proposal in his paper "Computing Machinery and Intelligence" that the question "Can machines think?" be replaced with the question "Can machines do what we (as thinking entities) can do?"

Types of Problems and Tasks

Machine learning tasks are typically classified into three broad categories, depending on the nature of the learning "signal" or "feedback" available to a learning system. These are

- Supervised learning: The computer is presented with example inputs and their desired outputs, given by a "teacher", and the goal is to learn a general rule that maps inputs to outputs.

- Unsupervised learning: No labels are given to the learning algorithm, leaving it on its own to find structure in its input. Unsupervised learning can be a goal in itself (discovering hidden patterns in data) or a means towards an end (feature learning).

- Reinforcement learning: A computer program interacts with a dynamic environment in which it must perform a certain goal (such as driving a vehicle), without a teacher explicitly telling it whether it has come close to its goal. Another example is learning to play a game by playing against an opponent.

Between supervised and unsupervised learning is semi-supervised learning, where the teacher gives an incomplete training signal: a training set with some (often many) of the target outputs missing. Transduction is a special case of this principle where the entire set of problem instances is known at learning time, except that part of the targets are missing.

Among other categories of machine learning problems, learning to learn learns its own inductive bias based on previous experience. Developmental learning, elaborated for robot learning, generates its own sequences (also called curriculum) of learning situations to cumulatively acquire repertoires of novel skills through autonomous self-exploration and social interaction with human teachers and using guidance mechanisms such as active learning, maturation, motor synergies, and imitation.

Another categorization of machine learning tasks arises when one considers the desired *output* of a machine-learned system:

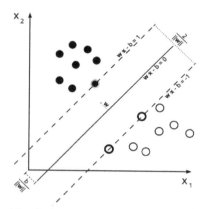

A support vector machine is a classifier that divides its input space into two regions, separated by a linear boundary. Here, it has learned to distinguish black and white circles.

- In classification, inputs are divided into two or more classes, and the learner must produce a model that assigns unseen inputs to one or more (multi-label classification) of these classes. This is typically tackled in a supervised way. Spam filtering is an example of classification, where the inputs are email (or other) messages and the classes are "spam" and "not spam".

- In regression, also a supervised problem, the outputs are continuous rather than discrete.

- In clustering, a set of inputs is to be divided into groups. Unlike in classification, the groups are not known beforehand, making this typically an unsupervised task.

- Density estimation finds the distribution of inputs in some space.

- Dimensionality reduction simplifies inputs by mapping them into a lower-dimensional space. Topic modeling is a related problem, where a program is given a list of human language documents and is tasked to find out which documents cover similar topics.

History and Relationships to other Fields

As a scientific endeavour, machine learning grew out of the quest for artificial intelligence. Already in the early days of AI as an academic discipline, some researchers were interested in having machines learn from data. They attempted to approach the problem with various symbolic methods, as well as what were then termed "neural networks"; these were mostly perceptrons and other models that were later found to be reinventions of the generalized linear models of statistics. Probabilistic reasoning was also employed, especially in automated medical diagnosis.

However, an increasing emphasis on the logical, knowledge-based approach caused a rift between AI and machine learning. Probabilistic systems were plagued by theoretical and practical problems of data acquisition and representation. By 1980, expert systems had come to dominate AI, and statistics was out of favor. Work on symbolic/knowledge-based learning did continue within AI, leading to inductive logic programming, but the more statistical line of research was now outside the field of AI proper, in pattern recognition and information retrieval. Neural networks research had been abandoned by AI and computer science around the same time. This line, too, was continued outside the AI/CS field, as "connectionism", by researchers from other disciplines including Hopfield, Rumelhart and Hinton. Their main success came in the mid-1980s with the reinvention of back-propagation.

Machine learning, reorganized as a separate field, started to flourish in the 1990s. The field changed its goal from achieving artificial intelligence to tackling solvable problems of a practical nature. It shifted focus away from the symbolic approaches it had inherited from AI, and toward methods and models borrowed from statistics and probability theory. It also benefited from the increasing availability of digitized information, and the possibility to distribute that via the Internet.

Machine learning and data mining often employ the same methods and overlap significantly. They can be roughly distinguished as follows:

- Machine learning focuses on prediction, based on *known* properties learned from the training data.

- Data mining focuses on the discovery of (previously) *unknown* properties in the data. This is the analysis step of Knowledge Discovery in Databases.

The two areas overlap in many ways: data mining uses many machine learning methods, but often with a slightly different goal in mind. On the other hand, machine learning also employs data mining methods as "unsupervised learning" or as a preprocessing step to improve learner accuracy. Much of the confusion between these two research communities (which do often have separate conferences and separate journals, ECML PKDD being a major exception) comes from the basic assumptions they work with: in machine learning, performance is usually evaluated with respect to the ability to *reproduce known* knowledge, while in Knowledge Discovery and Data Mining (KDD) the key task is the discovery of previously *unknown* knowledge. Evaluated with respect to known knowledge, an uninformed (unsupervised) method will easily be outperformed by supervised methods, while in a typical KDD task, supervised methods cannot be used due to the unavailability of training data.

Machine learning also has intimate ties to optimization: many learning problems are formulated as minimization of some loss function on a training set of examples. Loss functions express the discrepancy between the predictions of the model being trained and the actual problem instances (for example, in classification, one wants to assign a label to instances, and models are trained to correctly predict the pre-assigned labels of a set examples). The difference between the two fields arises from the goal of generalization: while optimization algorithms can minimize the loss on a training set, machine learning is concerned with minimizing the loss on unseen samples.

Relation to Statistics

Machine learning and statistics are closely related fields. According to Michael I. Jordan, the ideas of machine learning, from methodological principles to theoretical tools, have had a long pre-history in statistics. He also suggested the term data science as a placeholder to call the overall field.

Leo Breiman distinguished two statistical modelling paradigms: data model and algorithmic model, wherein 'algorithmic model' means more or less the machine learning algorithms like Random forest.

Some statisticians have adopted methods from machine learning, leading to a combined field that they call *statistical learning*.

Theory

A core objective of a learner is to generalize from its experience. Generalization in this context is the ability of a learning machine to perform accurately on new, unseen examples/tasks after having experienced a learning data set. The training examples come from some generally unknown probability distribution (considered representative of the space of occurrences) and the learner has to build a general model about this space that enables it to produce sufficiently accurate predictions in new cases.

The computational analysis of machine learning algorithms and their performance is a branch of theoretical computer science known as computational learning theory. Because training sets are finite and the future is uncertain, learning theory usually does not yield guarantees of the performance of algorithms. Instead, probabilistic bounds on the performance are quite common. The bias–variance decomposition is one way to quantify generalization error.

For the best performance in the context of generalization, the complexity of the hypothesis should match the complexity of the function underlying the data. If the hypothesis is less complex than the function, then the model has underfit the data. If the complexity of the model is increased in response, then the training error decreases. But if the hypothesis is too complex, then the model is subject to overfitting and generalization will be poorer.

In addition to performance bounds, computational learning theorists study the time complexity and feasibility of learning. In computational learning theory, a computation is considered feasible if it can be done in polynomial time. There are two kinds of time complexity results. Positive results show that a certain class of functions can be learned in polynomial time. Negative results show that certain classes cannot be learned in polynomial time.

Approaches

Decision Tree Learning

Decision tree learning uses a decision tree as a predictive model, which maps observations about an item to conclusions about the item's target value.

Association Rule Learning

Association rule learning is a method for discovering interesting relations between variables in large databases.

Artificial Neural Networks

An artificial neural network (ANN) learning algorithm, usually called "neural network" (NN), is a learning algorithm that is inspired by the structure and functional aspects of biological neural networks. Computations are structured in terms of an interconnected group of artificial neurons, processing information using a connectionist approach to computation. Modern neural networks are non-linear statistical data modeling tools. They are usually used to model complex relationships between inputs and outputs, to find patterns in data, or to capture the statistical structure in an unknown joint probability distribution between observed variables.

Deep Learning

Falling hardware prices and the development of GPUs for personal use in the last few years have contributed to the development of the concept of Deep learning which consists of multiple hidden layers in an artificial neural network. This approach tries to model the way the human brain processes light and sound into vision and hearing. Some successful applications of deep learning are computer vision and speech recognition.

Inductive Logic Programming

Inductive logic programming (ILP) is an approach to rule learning using logic programming as a uniform representation for input examples, background knowledge, and hypotheses. Given an encoding of the known background knowledge and a set of examples represented as a logical database of facts, an ILP system will derive a hypothesized logic program that entails all positive and no negative examples. Inductive programming is a related field that considers any kind of programming languages for representing hypotheses (and not only logic programming), such as functional programs.

Support Vector Machines

Support vector machines (SVMs) are a set of related supervised learning methods used for classification and regression. Given a set of training examples, each marked as belonging to one of two categories, an SVM training algorithm builds a model that predicts whether a new example falls into one category or the other.

Clustering

Cluster analysis is the assignment of a set of observations into subsets (called *clusters*) so that observations within the same cluster are similar according to some predesignated criterion or criteria, while observations drawn from different clusters are dissimilar. Different clustering techniques make different assumptions on the structure of the data, often defined by some *similarity metric* and evaluated for example by *internal compactness* (similarity between members of the same cluster) and *separation* between different clusters. Other methods are based on *estimated density* and *graph connectivity*. Clustering is a method of unsupervised learning, and a common technique for statistical data analysis.

Bayesian Networks

A Bayesian network, belief network or directed acyclic graphical model is a probabilistic graphical model that represents a set of random variables and their conditional independencies via a directed acyclic graph (DAG). For example, a Bayesian network could represent the probabilistic relationships between diseases and symptoms. Given symptoms, the network can be used to compute the probabilities of the presence of various diseases. Efficient algorithms exist that perform inference and learning.

Reinforcement Learning

Reinforcement learning is concerned with how an *agent* ought to take *actions* in an *environment* so as to maximize some notion of long-term *reward*. Reinforcement learning algorithms attempt to find a *policy* that maps *states* of the world to the actions the agent ought to take in those states. Reinforcement learning differs from the supervised learning problem in that correct input/output pairs are never presented, nor sub-optimal actions explicitly corrected.

Representation Learning

Several learning algorithms, mostly unsupervised learning algorithms, aim at discovering better representations of the inputs provided during training. Classical examples include principal components analysis and cluster analysis. Representation learning algorithms often attempt to preserve the information in their input but transform it in a way that makes it useful, often as a pre-processing step before performing classification or predictions, allowing to reconstruct the inputs coming from the unknown data generating distribution, while not being necessarily faithful for configurations that are implausible under that distribution.

Manifold learning algorithms attempt to do so under the constraint that the learned representation is low-dimensional. Sparse coding algorithms attempt to do so under the constraint that the learned representation is sparse (has many zeros). Multilinear subspace learning algorithms aim to learn low-dimensional representations directly from tensor representations for multidimensional data, without reshaping them into (high-dimensional) vectors. Deep learning algorithms discover multiple levels of representation, or a hierarchy of features, with higher-level, more abstract features defined in terms of (or generating) lower-level features. It has been argued that an intelligent machine is one that learns a representation that disentangles the underlying factors of variation that explain the observed data.

Similarity and Metric Learning

In this problem, the learning machine is given pairs of examples that are considered similar and pairs of less similar objects. It then needs to learn a similarity function (or a distance metric function) that can predict if new objects are similar. It is sometimes used in Recommendation systems.

Sparse Dictionary Learning

In this method, a datum is represented as a linear combination of basis functions, and the coefficients are assumed to be sparse. Let x be a d-dimensional datum, D be a d by n matrix, where each column of D represents a basis function. r is the coefficient to represent x using D. Mathematically, sparse dictionary learning means solving where r is sparse. Generally speaking, n is assumed to be larger than d to allow the freedom for a sparse representation.

Learning a dictionary along with sparse representations is strongly NP-hard and also difficult to solve approximately. A popular heuristic method for sparse dictionary learning is K-SVD.

Sparse dictionary learning has been applied in several contexts. In classification, the problem is to determine which classes a previously unseen datum belongs to. Suppose a dictionary for each class has already been built. Then a new datum is associated with the class such that it's best sparsely represented by the corresponding dictionary. Sparse dictionary learning has also been applied in image de-noising. The key idea is that a clean image patch can be sparsely represented by an image dictionary, but the noise cannot.

Genetic Algorithms

A genetic algorithm (GA) is a search heuristic that mimics the process of natural selection, and uses methods such as mutation and crossover to generate new genotype in the hope of finding good solutions to a given problem. In machine learning, genetic algorithms found some uses in the 1980s and 1990s. Vice versa, machine learning techniques have been used to improve the performance of genetic and evolutionary algorithms.

Applications

Applications for machine learning include:

- Adaptive websites
- Affective computing
- Bioinformatics
- Brain-machine interfaces
- Cheminformatics
- Classifying DNA sequences
- Computational anatomy
- Computer vision, including object recognition
- Detecting credit card fraud
- Game playing
- Information retrieval
- Internet fraud detection
- Marketing
- Machine perception
- Medical diagnosis
- Natural language processing
- Natural language understanding
- Optimization and metaheuristic
- Online advertising
- Recommender systems
- Robot locomotion
- Search engines
- Sentiment analysis (or opinion mining)

- Sequence mining
- Software engineering
- Speech and handwriting recognition
- Stock market analysis
- Structural health monitoring
- Syntactic pattern recognition
- Economics

In 2006, the online movie company Netflix held the first "Netflix Prize" competition to find a program to better predict user preferences and improve the accuracy on its existing Cinematch movie recommendation algorithm by at least 10%. A joint team made up of researchers from AT&T Labs-Research in collaboration with the teams Big Chaos and Pragmatic Theory built an ensemble model to win the Grand Prize in 2009 for $1 million. Shortly after the prize was awarded, Netflix realized that viewers' ratings were not the best indicators of their viewing patterns ("everything is a recommendation") and they changed their recommendation engine accordingly.

In 2010 The Wall Street Journal wrote about money management firm Rebellion Research's use of machine learning to predict economic movements. The article describes Rebellion Research's prediction of the financial crisis and economic recovery.

In 2014 it has been reported that a machine learning algorithm has been applied in Art History to study fine art paintings, and that it may have revealed previously unrecognized influences between artists.

Ethics

Machine Learning poses a host of ethical questions. Systems which are trained on datasets collected with biases may exhibit these biases upon use, thus digitizing cultural prejudices such as institutional racism and classism. Responsible collection of data thus is a critical part of machine learning.

Computational Learning Theory

In computer science, computational learning theory (or just learning theory) is a subfield of Artificial Intelligence devoted to studying the design and analysis of machine learning algorithms.

Overview

Theoretical results in machine learning mainly deal with a type of inductive learning called supervised learning. In supervised learning, an algorithm is given samples that are labeled in some useful way. For example, the samples might be descriptions of mushrooms, and the labels could be whether or not the mushrooms are edible. The algorithm takes these previously labeled samples and uses them to induce a classifier. This classifier is a function that assigns labels to samples including the samples that have never been previously seen by the algorithm. The goal of the supervised learning algorithm is to optimize some measure of performance such as minimizing the number of mistakes made on new samples.

In addition to performance bounds, computational learning theory studies the time complexity and feasibility of learning. In computational learning theory, a computation is considered feasible if it can be done in polynomial time. There are two kinds of time complexity results:

- Positive results – Showing that a certain class of functions is learnable in polynomial time.

- Negative results – Showing that certain classes cannot be learned in polynomial time.

Negative results often rely on commonly believed, but yet unproven assumptions, such as:

- Computational complexity – P ≠ NP (the P versus NP problem);

- Cryptographic – One-way functions exist.

There are several different approaches to computational learning theory. These differences are based on making assumptions about the inference principles used to generalize from limited data. This includes different definitions of probability and different assumptions on the generation of samples. The different approaches include:

- Exact learning, proposed by Dana Angluin;

- Probably approximately correct learning (PAC learning), proposed by Leslie Valiant;

- VC theory, proposed by Vladimir Vapnik and Alexey Chervonenkis;

- Bayesian inference;

- Algorithmic learning theory, from the work of E. Mark Gold;

- Online machine learning, from the work of Nick Littlestone.

Computational learning theory has led to several practical algorithms. For example,

PAC theory inspired boosting, VC theory led to support vector machines, and Bayesian inference led to belief networks (by Judea Pearl).

Computational Anatomy

Computational anatomy is a discipline within medical imaging focusing on the study of anatomical shape and form at the morphome scale of morphology. It involves the development and application of computational, mathematical and data-analytical methods for modeling and simulation of biological structures. The field is broadly defined and includes foundations in anatomy, applied mathematics and pure mathematics, machine learning, computational mechanics, computational science, medical imaging, neuroscience, physics, probability, and statistics; it also has strong connections with fluid mechanics and geometric mechanics. Additionally, it complements newer, interdisciplinary fields like bioinformatics and neuroinformatics in the sense that its interpretation uses metadata derived from the original sensor imaging modalities (of which Magnetic Resonance Imaging is one example). It focuses on the anatomical structures being imaged, rather than the medical imaging devices. It is similar in spirit to the history of Computational linguistics, a discipline that focuses on the linguistic structures rather than the detectors acting as the transmission and communication medium(s).

In Computational anatomy, the diffeomorphism group for coordinate transformations is generated via the Lagrangian and Eulerian velocities of flow in $\mathbb{R}^3$. The flows between coordinates in Computational anatomy are constrained to be geodesic flows satisfying the principle of least action for the Kinetic energy of the flow defined via a Sobolev smoothness norm with more than two finite square-integrable derivatives for each component of the velocity of flow. This, in turn, guarantees that the flows in $\mathbb{R}^3$ are diffeomorphisms; it also implies that the diffeomorphic shape momentum in Computational anatomy, which satisfies the Euler-Lagrange equation for geodesics, is determined by its velocity and spatial derivatives. This separates the discipline from the case of incompressible fluids for which momentum is a pointwise function of velocity. Computational anatomy intersects the study of Riemannian manifolds and nonlinear global analysis, where groups of diffeomorphisms are the central focus. Emerging high-dimensional theories of shape are central to many studies in Computational anatomy, as are questions emerging from the fledgling field of shape statistics. The metric structures in Computational anatomy are related in spirit to morphometrics, with the distinction that Computational anatomy focuses on an infinite-dimensional space of coordinate systems transformed by a diffeomorphism, hence the central use of the terminology diffeomorphometry, the metric space study of coordinate systems via diffeomorphisms.

Genesis

At Computational anatomy's heart is the comparison of shape by recognizing in one shape the other, connecting it to D'Arcy Wentworth Thompson's developments On Growth and Form which has led to scientific explanations of morphogenesis, the process by which patterns are formed in Biology. Albrecht Durer's Four Books on Human Proportion were arguably the earliest works on Computational anatomy. The Grenander abstraction within the setting of deformable templates uses group actions as the comparison mechanism; the central group action of Computational anatomy is diffeomorphic action. The efforts of Noam Chomsky in his pioneering of Computational Linguistics inspired the original formulation of Computational anatomy as a generative model of shape and form from exemplars acted upon via transformations.

Due to the focus on medical imaging technologies such as magnetic resonance imaging (MRI), Computational anatomy has emerged as a subfield of medical imaging and bioengineering for extracting anatomical coordinate systems at the morphome scale in 3D. The spirit of this discipline shares strong overlap with areas such as computer vision and kinematics of rigid bodies, where objects are studied by analysing the groups responsible for the movement in question. Computational anatomy departs from computer vision with its focus on rigid motions, as the infinite-dimensional diffeomorphism group is central to the analysis of Biological shapes. It is a branch of the image analysis and pattern theory school at Brown University pioneered by Ulf Grenander. In Grenander's general Metric Pattern Theory, making spaces of patterns into a metric space is one of the fundamental operations since being able to cluster and recognize anatomical configurations often requires a metric of close and far between shapes. The diffeomorphometry metric of Computational anatomy measures how far two diffeomorphic changes of coordinates are from each other, which in turn induces a metric on the shapes and images indexed to them. The models of metric pattern theory, in particular group action on the orbit of shapes and forms is a central tool to the formal definitions in Computational anatomy.

History

Computational anatomy is the study of shape and form at the morphome or gross anatomy millimeter, or morphology scale, focusing on the study of sub-manifolds of $\mathbb{R}^3$ points, curves surfaces and subvolumes of human anatomy. An early modern computational neuro-anatomist was David Van Essen performing some of the early physical unfoldings of the human brain based on printing of a human cortex and cutting. Jean Talairach's publication of Tailarach coordinates is an important milestone at the morphome scale demonstrating the fundamental basis of local coordinate systems in studying neuroanatomy and therefore the clear link to charts of differential geometry. Concurrently, virtual mapping in Computational anatomy across high resolution dense image coordinates was already happening in Ruzena Bajcy's and Fred Bookstein's earliest developments based on Computed axial tomography and Magnetic resonance imagery. The earliest introduction of the use of flows of diffeomorphisms for transformation of coordinate systems in image analysis and medical imaging was by Christensen, Joshi, Miller, and Rabbitt.

The first formalization of Computational Anatomy as an orbit of exemplar templates under diffeomorphism group action was in the original lecture given by Grenander and Miller with that title in May 1997 at the 50th Anniversary of the Division of Applied Mathematics at Brown University, and subsequent publication. This was the basis for the strong departure from much of the previous work on advanced methods for spatial normalization and image registration which were historically built on notions of addition and basis expansion. The structure preserving transformations central to the modern field of Computational Anatomy, homeomorphisms and diffeomorphisms carry smooth submanifolds smoothly. They are generated via Lagrangian and Eulerian flows which satisfy a law of composition of functions forming the group property, but are not additive.

The original model of Computational anatomy was as the triple, $(\mathcal{G}, \mathcal{M}, \mathcal{P})$, the group $g \in \mathcal{G}$, the orbit of shapes and forms $m \in \mathcal{M}$, and the probability laws P which encode the variations of the objects in the orbit. The template or collection of templates are elements in the orbit $m_{\text{temp}} \in \mathcal{M}$ of shapes.

The Lagrangian and Hamiltonian formulations of the equations of motion of Computational Anatomy took off post 1997 with several pivotal meetings including the 1997 Luminy meeting organized by the Azencott school at Ecole-Normale Cachan on the "Mathematics of Shape Recognition" and the 1998 Trimestre at Institute Henri Poincaré organized by David Mumford "Questions Mathématiques en Traitement du Signal et de l'Image" which catalyzed the Hopkins-Brown-ENS Cachan groups and subsequent developments and connections of Computational anatomy to developments in global analysis.

The developments in Computational Anatomy included the establishment of the Sobelev smoothness conditions on the diffeomorphometry metric to insure existence of solutions of variational problems in the space of diffeomorphisms, the derivation of the Euler-Lagrange equations characterizing geodesics through the group and associated conservation laws, the demonstration of the metric properties of the right invariant metric, the demonstration that the Euler-Lagrange equations have a well-posed initial value problem with unique solutions for all time, and with the first results on sectional curvatures for the diffeomorphometry metric in landmarked spaces. Following the Los Alamos meeting in 2002, Joshi's original large deformation singular *Landmark* solutions in Computational anatomy were connected to peaked *Solitons* or *Peakons* as solutions for the Camassa-Holm equation. Subsequently connections were made between Computational anatomy's Euler-Lagrange equations for momentum densities for the right-invariant metric satisfying Sobolev smoothness to Vladimir Arnold's characterization of the Euler equation for incompressible flows as describing geodesics in the group of volume preserving diffeomorphisms. The first algorithms, generally termed LDDMM for large deformation diffeomorphic mapping for computing connections between landmarks in volumes and spherical manifolds, curves, currents and surfaces, volumes, tensors, varifolds, and time-series have followed.

These contributions of Computational anatomy to the global analysis associated to the infinite dimensional manifolds of subgroups of the diffeomorphism group is far from trivial. The original idea of doing differential geometry, curvature and geodesics on infinite dimensional manifolds goes back to Bernhard Riemann's Habilitation (Ueber die Hypothesen, welche der Geometrie zu Grunde liegen); the key modern book laying the foundations of such ideas in global analysis are from Michor.

The applications within Medical Imaging of Computational Anatomy continued to flourish after two organized meetings at the Institute for Pure and Applied Mathematics conferences at University of California, Los Angeles. Computational anatomy has been useful in creating accurate models of the atrophy of the human brain at the morphome scale, as well as Cardiac templates, as well as in modeling biological systems. Since the late 1990s, computational anatomy has become an important part of developing emerging technologies for the field of medical imaging. Digital atlases are a fundamental part of modern Medical-school education and in neuroimaging research at the morphome scale. Atlas based methods and virtual textbooks which accommodate variations as in deformable templates are at the center of many neuro-image analysis platforms including Freesurfer, FSL, MRIStudio, SPM. Diffeomorphic registration, introduced in the 90's, is now an important player with existing codes bases organized around ANTS, DARTEL, DEMONS, LDDMM, StationaryLDDMM are examples of actively used computational codes for constructing correspondences between coordinate systems based on sparse features and dense images. Voxel-based morphometry(VBM) is an important technology built on many of these principles.

The Deformable Template Orbit Model of Computational Anatomy

The model of human anatomy is a deformable template, an orbit of exemplars under group action. Deformable template models have been central to Grenander's Metric Pattern theory, accounting for typicality via templates, and accounting for variability via transformation of the template. An orbit under group action as the representation of the deformable template is a classic formulation from differential geometry. The space of shapes are denoted $m \in \mathcal{M}$, with the group $(\mathcal{G}, \circ)$ with law of composition $\circ$; the action of the group on shapes is denoted $g \cdot m$;, where the action of the group $g \cdot m \in \mathcal{M}, m \in \mathcal{M}$ is defined to satisfy

$$(g \circ g') \cdot m = g \cdot (g' \cdot m) \in \mathcal{M}.$$

The orbit $\mathcal{M}$ of the template becomes the space of all shapes, $\mathcal{M} \doteq \{m = g \cdot m_{\text{temp}}, g \in \mathcal{G}\}$, being homogenous under the action of the elements of .

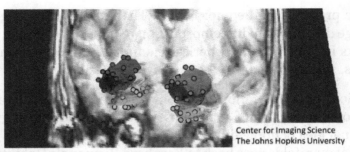

Center for Imaging Science
The Johns Hopkins University

Figure depicting three medial temporal lobe structures amgydala, entorhinal cortex and hippocampus
with fiducial landmarks depicted as well embedded in the MRI background.

The orbit model of computational anatomy is an abstract algebra - to be compared to
linear algebra- since the groups act nonlinearly on the shapes. This is a generalization
of the classical models of linear algebra, in which the set of finite dimensional $\mathbb{R}^n$ vectors are replaced by the finite-dimensional anatomical submanifolds (points, curves,
surfaces and volumes) and images of them, and the $n \times n$ matrices of linear algebra are
replaced by coordinate transformations based on linear and affine groups and the more
general high-dimensional diffeomorphism groups.

Shapes and Forms

The central objects are shapes or forms in Computational anatomy, one set of examples
being the 0,1,2,3-dimensional submanifolds of $\mathbb{R}^3$, a second set of examples being images generated via medical imaging such as via magnetic resonance imaging (MRI) and
functional magnetic resonance imaging.

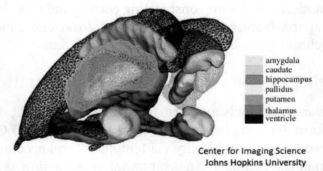

amygdala
caudate
hippocampus
pallidus
putamen
thalamus
ventricle

Center for Imaging Science
Johns Hopkins University

Triangulated mesh surfaces depicting subcortical structures amygdala, hippocampus, thalamus, caudate,

putamen, ventricles.The shapes are denoted $m(u), u \in U \subset \mathbb{R}^1 \to \mathbb{R}^2$ represented as triangulated
meshes.

The 0-dimensional manifolds are landmarks or fiducial points; 1-dimensional manifolds are curves such as sulcul and gyral curves in the brain; 2-dimensional manifolds
correspond to boundaries of substructures in anatomy such as the subcortical structures of the midbrain or the gyral surface of the neocortex; subvolumes correspond to
subregions of the human body, the heart, the thalamus, the kidney.

The landmarks, denoted as $X \doteq \{x_1, \ldots, x_n\} \subset \mathbb{R}^3 \in \mathcal{M}$ are a collections of points with no other structure, delineating important fiducials within human shape and form.

The sub-manifold shapes such as surfaces are denoted as $X \subset \mathbb{R}^3 \in \mathcal{M}$, , collections of points modeled as parametrized by a local chart or immersion $m : U \subset \mathbb{R}^{1,2} \to \mathbb{R}^3$, $m(u), u \in U$.

The images in Computational anatomy such as MR images or DTI images are denoted $I \in \mathcal{M}$, and are dense functions $I(x), x \in X \subset \mathbb{R}^{1,2,3}$ are scalars, vectors, and matrices

Groups and Group Actions

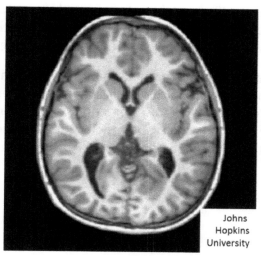

Johns
Hopkins
University

Showing an MRI section through a 3D brain representing a scalar image $I(x), x \in \mathbb{R}^2$ based on T1-weighting.

Groups and group actions are familiar to the Engineering community with the universal popularization and standardization of linear algebra as a basic model for analyzing signals and systems in mechanical engineering, electrical engineering and applied mathematics. In linear algebra the matrix groups (matrices with inverses) are the central structure, with group action defined by the usual definition of A as an $n \times n$ matrix, acting on $x \in \mathbb{R}^n$ as $n \times 1$ vectors; the orbit in linear-algebra is the set of n-vectors given by $y = A \cdot x \in \mathbb{R}^n$, which is a group action of the matrices through the orbit of $\mathbb{R}^n$.

The central group in Computational anatomy defined on volumes in $\mathbb{R}^3$ are the diffeomorphisms $\mathcal{G} \doteq Diff$ which are mappings with 3-components $\phi(\cdot) = (\phi_1(\cdot), \phi_2(\cdot), \phi_3(\cdot))$, law of composition of functions $\phi \circ \phi'(\cdot) \doteq \phi(\phi'(\cdot))$, with inverse $\phi \circ \phi^{-1}(\cdot) = \phi(\phi^{-1}(\cdot)) = id$.

Most popular are scalar images, $I(x), x \in \mathbb{R}^3$, with action on the right via the inverse.

$$\phi \cdot I(x) = I \circ \phi^{-1}(x), x \in \mathbb{R}^3.$$

For sub-manifolds $X \subset \mathbb{R}^3 \in \mathcal{M}$, parametrized by a chart or immersion $m(u), u \in U$, the diffeomorphic action the flow of the position

$$\phi \cdot m(u) \doteq \phi^\circ m(u), u \in U.$$

Several group actions in computational anatomy have been defined.

Lagrangian and Eulerian Flows for Generating Diffeomorphisms

For the study of rigid body kinematics, the low-dimensional matrix Lie groups have been the central focus. The matrix groups are low-dimensional mappings, which are diffeomorphisms that provide one-to-one correspondences between coordinate systems, with a smooth inverse. The matrix group of rotations and scales can be generated via a closed form finite-dimensional matrices which are solution of simple ordinary differential equations with solutions given by the matrix exponential.

For the study of deformable shape in Computational anatomy, a more general diffeomorphism group has been the group of choice, which is the infinite dimensional analogue. The high-dimensional differeomorphism groups used in Computational Anatomy are generated via smooth flows $\phi_t, t \in [0,1]$ which satisfy the Lagrangian and Eulerian specification of the flow fieldssas first introduced in., satisfying the ordinary differential equation:

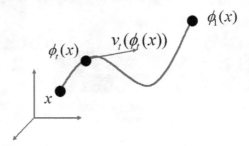

with $v \doteq (v_1, v_2, v_3)$ the vector fields on $\mathbb{R}^3$ termed the Eulerian velocity of the particles at position ϕ of the flow. The vector fields are functions in a function space, modelled as a smooth Hilbert space of high-dimension, with the Jacobian of the flow $D\phi \doteq (\frac{\partial \phi_i}{\partial x_j})$ a high-dimensional field in a function space as well, rather than a low-dimensional matrix as in the matrix groups. Flows were first introduced for large deformations in image matching; $\dot{\phi}_t(x)$ is the instantaneous velocity of particle x at time t.

The inverse $\phi_t^{-1}, t \in [0,1]$ required for the group is defined on the Eulerian vector-field with advective inverse flow

(Inverse Transport flow)

$$\frac{d}{dt} \phi_t^{-1} = -(D\phi_t^{-1})v_t, \phi_0^{-1} = id .$$

The Diffeomorphism Group of Computational Anatomy

The group of diffeomorphisms is very big. To ensure smooth flows of diffeomorphisms avoiding shock-like solutions for the inverse, the vector fields must be at least 1-time continuously differentiable in space. For diffeomorphisms on $\mathbb{R}^3$, vector fields are modelled as elements of the Hilbert space $(V, \|\cdot\|_V)$ using the Sobolev embedding theorems so that each element $v_i \in H_0^3, i = 1, 2, 3$, has 3-square-integrable derivatives, thusly embedding in 1-time continuously differentiable functions.

The diffeomorphism group are flows with vector fields absolutely integrable in Sobolev norm:

$$Diff_V \doteq \{\varphi = \phi_1 : \dot{\phi}_t = v_t \circ \phi_t, \phi_0 = id, \int_0^1 \|v_t\|_V dt < \infty\}. \qquad \textit{(Diffeomorphism Group)}$$

The Sobolev Smoothness Condition on Vector Fields as Modeled in a Reproducing Kernel Hilbert Space

The modelling approach used in Computational anatomy enforces a continuous differentiability condition on the vector fields by modelling the space of vector fields $(V, \|\cdot\|_V)$ as a reproducing kernel Hilbert space (RKHS), with the norm defined by a 1-1, differential operator $A : V \rightarrow V^*$, Green's inverse $K = A^{-1}$. The norm of the Hilbert space is induced by the differential operator. For $\sigma(v) \doteq Av \in V^*$ a generalized function or distribution, define the linear form as $(\sigma \mid w) \quad \int \sum w_i(x)\sigma_i(dx)$. This determines the norm on $(V, \|\cdot\|_V)$ according to

$$\langle v, w \rangle_V \doteq (Av \mid w), \|v\|_V^2 \doteq (Av \mid v), v, w \in V.$$

Since A is a differential operator, finiteness of the norm-square $(Av \mid v) < \infty$ includes derivatives from the differential operator implying smoothness of the vector fields. The Sobolev embedding theorem arguments were made in demonstrating that 1-continuous derivative is required for smooth flows.

For proper choice of A then $(V, \|\cdot\|_V)$ is an RKHS with the operator $K = A^{-1} : V^* \rightarrow V$ termed the Green's operator generated from the Green's function (scalar case) for the vector field case. The Green's kernels associated to the differential operator smooths since the kernel $k(\cdot, \cdot)$ is continuously differentiable in both variables implying

$$K\sigma(x)_i \doteq \sum_j \int_{\mathbb{R}^3} k_{ij}(x, y)\sigma_j(dy)$$

When $\sigma \doteq \mu dx$, a vector density, $(\sigma \mid v) \doteq \int v \cdot \mu dx..$

Diffeomorphometry: The Metric Space of Shapes and Forms

The study of metrics on groups of diffeomorphisms and the study of metrics between manifolds and surfaces has been an area of significant investigation. In Computational anatomy, the diffeomorphometry metric measures how close and far two shapes or images are from each other. Informally, the metric length is the shortest length of the flow which carries one coordinate system into the other.

Oftentimes, the familiar Euclidean metric is not directly applicable because the patterns of shapes and images don't form a vector space. In the Riemannian orbit model of

Computationalanatomy,diffeomorphismsactingontheforms $\phi \cdot m \in \mathcal{M}, \phi \in Diff_V, m \in \mathcal{M}$

don't act linearly. There are many ways to define metrics, and for the sets associated to shapes the Hausdorff metric is another. The method we use to induce the Riemannian metric is used to induce the metric on the orbit of shapes by defining it in terms of the metric length between diffeomorphic coordinate system transformations of the flows. Measuring the lengths of the geodesic flow between coordinates systems in the orbit of shapes is called diffeomorphometry.

The Right-invariant Metric on Diffeomorphisms

Define the distance on the group of diffeomorphisms

$$: d_{Diff_V}(\psi, \varphi) = \inf_{v_t} \left(\int_0^1 (Av_t \mid v_t) dt : \phi_0 = \psi, \phi_1 = \varphi, \dot{\phi}_t = v_t \circ \phi_t \right)^{1/2} ; \qquad (metric\text{-}diffeomorphisms)$$

this is the right-invariant metric of diffeomorphometry, invariant to reparameterization of space since for all $\phi \in Diff_V$,

$$d_{Diff_V}(\psi, \varphi) = d_{Diff_V}(\psi \circ \phi, \varphi \circ \phi).$$

The Metric on Shapes and Forms

The distance on shapes and forms, $d_{\mathcal{M}} : \mathcal{M} \times \mathcal{M} \to \mathbb{R}^+$,

$$: d_{\mathcal{M}}(m, n) = \inf_{\phi \in Diff_V : \phi \cdot m = n} d_{Diff_V}(id, \phi) ; \qquad (metric\text{-}shapes\text{-}forms)$$

the images are denoted with the orbit as $I \in \mathcal{I}$ and metric $, d_{\mathcal{I}}$.

The Action Integral for Hamilton's Principle on Diffeomorphic Flows

In classical mechanics the evolution of physical systems is described by solutions to the Euler–Lagrange equations associated to the Least-action principle of Hamilton. This is a standard way, for example of obtaining Newton's laws of motion of free particles.

More generally, the Euler-Lagrange equations can be derived for systems of generalized coordinates. The Euler-Lagrange equation in Computational anatomy describes the geodesic shortest path flows between coordinate systems of the diffeomorphism metric. In Computational anatomy the generalized coordinates are the flow of the diffeomorphism and its Lagrangian velocity $\phi, \dot{\phi}$, the two related via the Eulerian velocity $v \doteq \dot{\phi} \circ \phi^{-1}$. Hamilton's principle for generating the Euler-Lagrange equation requires the action integral on the Lagrangian given by

$$J(\phi) \doteq \int_0^1 L(\phi_t, \dot{\phi}_t)dt \; ; \qquad \text{(Hamiltonian-Integrated-Lagrangian)}$$

the Lagrangian is given by the kinetic energy:

$$L(\phi_t, \dot{\phi}_t) \doteq \frac{1}{2}(A(\dot{\phi}_t \circ \phi_t^{-1}) \mid \dot{\phi}_t \circ \phi_t^{-1}) = \frac{1}{2}(Av_t \mid v_t) . \qquad \text{(Lagrangian-Kinetic-Energy)}$$

Diffeomorphic or Eulerian Shape Momentum

In computational anatomy, Av was first called the Eulerian or diffeomorphic shape momentum since when integrated against Eulerian velocity A gives energy density, and since there is a conservation of diffeomorphic shape momentum which holds. The operator A is the generalized moment of inertia or inertial operator.

The Euler–lagrange Equation on Shape Momentum for Geodesics on the Group of Diffeomorphisms

Classical calculation of the Euler-Lagrange equation from Hamilton's principle requires the perturbation of the Lagrangian on the vector field in the kinetic energy with respect to first order perturbation of the flow. This requires adjustment by the Lie bracket of vector field, given by operator $ad_v : w \in V \mapsto V$ which involves the Jacobian given by

$$ad_v[w] \doteq [v, w] \doteq (Dv)w - (Dw)v \in V.$$

Defining the adjoint $ad_v^* : V^* \to V^*$, then the first order variation gives the Eulerian shape momentum $Av \in V^*$ satisfying the generalized equation:

$$\frac{d}{dt}Av_t + ad_{v_t}^*(Av_t) = 0 , t \in [0,1] ; \qquad \text{(EL-General)}$$

meaning for all smooth $w \in V$,

$$\left(\frac{d}{dt}Av_t + ad_{v_t}^*(Av_t) \mid w\right) = (\frac{d}{dt}Av_t \mid w) + (Av_t \mid (Dv_t)w - (Dw)v_t) = 0.$$

Computational anatomy is the study of the motions of submanifolds, points, curves, surfaces and volumes. Momentum associated to points, curves and surfaces are all singular, implying the momentum is concentrated on subsets of $\mathbb{R}^3$ which are dimension ≤ 2 in Lebesgue measure. In such cases, the energy is still well defined $(Av_t \mid v_t)$ since although Av_t is a generalized function, the vector fields are smooth and the Eulerian momentum is understood via its action on smooth functions. The perfect illustration of this is even when it is a superposition of delta-diracs, the velocity of the coordinates in the entire volume move smoothly.The Euler-Lagrange equation (EL-General) on diffeomorphisms for generalized functions $Av \in V^*$ was derived in. In Riemannian Metric and Lie-Bracket Interpretation of the Euler-Lagrange Equation on Geodesics derivations are provided in terms of the adjoint operator and the Lie bracket for the group of diffeomorphisms. It has come to be called EPDiff equation for diffeomorphisms connecting to the Euler-Poincare method having been studied in the context of the inertial operator $A = identity$ for incompressible, divergence free, fluids.

Diffeomorphic Shape Momentum: a Classical Vector Function

For the momentum density case $(Av_t \mid w) = \int_X \mu_t \cdot w dx$, then Euler–Lagrange equation has a classical solution:

$$\frac{d}{dt}\mu_t + (Dv_t)^T \mu_t + (D\mu_t)v_t + (\nabla \cdot v)\mu_t = 0, t \in [0,1]. \qquad (EL\text{-}Classic)$$

The Euler-Lagrange equation on diffeomorphisms, classically defined for momentum densities first appeared in for medical image analysis.

Riemannian Exponential (Geodesic Positioning) and Riemannian Logarithm (Geodesic Coordinates)

Global positioning systems based on systems of satellites provides a spatial navigation sytstem on the globe allowing electronic receivers to determine their location in the 3-dimensional coordinate system of longitude, latitude, and altitude to within meter scale.

In Medical imaging and Computational anatomy, positioning and coordinatizing shapes are fundamental operations; the system for positioning anatomical coordinates and shapes built on the metric and the Euler-Lagrange equation a geodesic positioning system as first explicated in Miller Trouve and Younes. Solving the geodesic from the initial condition v_0 is termed the Riemannian-exponential, a mapping $Exp_{id}(\cdot) : V \to Diff_V$ at identity to the group.

The Riemannian exponential satisfies $Exp_{id}(v_0) = \phi_1$ for initial condition $\dot{\phi}_0 = v_0$, vector field dynamics $\dot{\phi}_t = v_t \circ \phi_t, t \in [0,1]$,

- for classical equation diffeomorphic shape momentum

$$(Av_t \mid w) = \int_X Av_t \cdot w dx, \ Av \in V, \text{ then}$$

$$\frac{d}{dt} Av_t + (Dv_t)^T Av_t + (DAv_t)v_t + (\nabla \cdot v)Av_t = 0 ;$$

- for generalized equation, then $Av \in V^*, w \in V$,

$$(\frac{d}{dt} Av_t \mid w) + (Av_t \mid (Dv_t)w - (Dw)v_t) = 0.$$

Computing the flow v_0 onto coordinates Riemannian logarithm, mapping $Log_{id}(\cdot): Diff_V \to V$ at identity from φ to vector field $v_0 \in V$;

$$Log_{id}(\varphi) = v_0 \text{ initial condition of EL geodesic } \dot{\phi}_0 = v_0, \phi_0 = id, \phi_1 = \varphi.$$

Extended to the entire group they become

$$\phi = Exp_\varphi(v_0 \circ \varphi) \doteq Exp_{id}(v_0) \circ \varphi \ ; \ Log_\varphi(\phi) \doteq Log_{id}(\phi \circ \varphi^{-1}) \circ \varphi.$$

These are inverses of each other for unique solutions of Logarithm; the first is called geodesic positioning, the latter geodesic coordinates (see Exponential map, Riemannian geometry for the finite dimensional version).The geodesic metric is a local flattening of the Riemannian coordinate system (see figure).

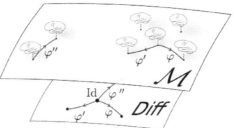

Showing metric local flattening of coordinatized manifolds of shapes and forms. The local metric is given by the norm of the vector field $Exp_{id}(v_0) \cdot m$ of the geodesic mapping

Hamiltonian Formulation of Computational Anatomy

In Computational anatomy the diffeomorphisms are used to push the coordinate systems, and the vector fields are used as the control within the anatomical orbit or morphological space. The model is that of a dynamical system, the flow of coordinates

$t \mapsto \phi_t \in \mathit{Diff}_V$ and the control the vector field $t \mapsto v_t \in V$ related via $\dot{\phi}_t = v_t \cdot \phi_t, \phi_0 = id$. The Hamiltonian view reparameterizes the momentum distribution $Av \in V^*$ in terms of the *conjugate momentum or canonical momentum,* introduced as a Lagrange multiplier $p : \dot{\phi} \mapsto (p \mid \dot{\phi})$ constraining the Lagrangian velocity $\dot{\phi}_t = v_t \circ \phi_t$ .accordingly:

$$H(\phi_t, p_t, v_t) = (p_t \mid v_t \circ \phi_t) - \frac{1}{2}(Av_t \mid v_t).$$

This function is the extended Hamiltonian. The Pontryagin maximum principle gives the optimizing vector field which determines the geodesic flow satisfying $\dot{\phi}_t = v_t \circ \phi_t, \phi_0 = id$, as well as the reduced Hamiltonian

$$H(\phi_t, p_t) \doteq \max_v H(\phi_t, p_t, v).$$

The Lagrange multiplier in its action as a linear form has its own inner product of the canonical momentum acting on the velocity of the flow which is dependent on the shape, e.g. for landmarks a sum, for surfaces a surface integral, and. for volumes it is a volume integral with respect to dx on $\mathbb{R}^3$. In all cases the Greens kernels carry weights which are the canonical momentum evolving according to an ordinary differential equation which corresponds to EL but is the geodesic reparameterization in canonical momentum. The optimizing vector field is given by

$$v_t \doteq \arg max_v H(\phi_t, p_t, v)$$

with dynamics of canonical momentum reparameterizing the vector field along the geodesic

$$\begin{cases} \dot{\phi}_t = \dfrac{\partial H(\phi_t, p_t)}{\partial p} \\[2mm] \dot{p}_t = -\dfrac{\partial H(\phi_t, p_t)}{\partial \phi} \end{cases} \qquad \text{(Hamiltonian-Dynamics)}$$

Stationarity of the Hamiltonian and Kinetic Energy Along Euler–Lagrange

Whereas the vector fields are extended across the entire background space of $\mathbb{R}^3$, the geodesic flows associated to the submanifolds has Eulerian shape momentum which evolves as a generalized function $Av_t \in V^*$ concentrated to the submanifolds. For land

marks the geodesics have Eulerian shape momentum which are a superposition of delta distributions travelling with the finite numbers of particles; the diffeomorphic flow of coordinates have velocities in the range of weighted Green's Kernels. For surfaces, the momentum is a surface integral of delta distributions travelling with the surface.

The geodesics connecting coordinate systems satisfying EL-General have stationarity of the Lagrangian. The Hamiltonian is given by the extremum along the path $t \in [0,1]$, $H(\phi, p) = \max_v H(\phi, p, v)$, equalling the Lagrangian-Kinetic-Energy and is stationary along EL-General. Defining the geodesic velocity at the identity

$v_0 = \arg\max_v H(\phi_0, p_0, v)$, then along the geodesic

$$H(\phi_t, p_t) = H(\phi_0, p_0) = \frac{1}{2}(p_0 \mid v_0) = \frac{1}{2}(Av_0 \mid v_0) = \frac{1}{2}(Av_t \mid v_t) \quad \text{(Hamiltonian-Geodesics)}$$

The stationarity of the Hamiltonian demonstrates the interpretation of the Lagrange multiplier as momentum; integrated against velocity $\dot\phi$ gives energy density. The canonical momentum has many names. In optimal control, the flows ϕ is interpreted as the state, and p is interpreted as conjugate state, or conjugate momentum. The geodesi of EL implies specification of the vector fields v_0 or Eulerian momentum Av_0 at $t = 0$, or specification of canonical momentum p_0 determines the flow.

The Metric on Geodesic Flows of Landmarks, Surfaces, and Volumes within the Orbit

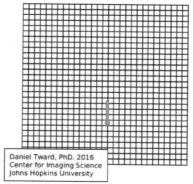

Daniel Tward, PhD. 2016
Center for Imaging Science
Johns Hopkins University

Illustration of geodesic flow for one landmark, demonstrating diffeomorphic motion of background space. Red arrow shows $p_0(1)$, blue curve shows $\varphi_t(x_1)$, φ_t black grid shows

In Computational anatomy the submanifolds are pointsets, curves, surfaces and subvolumes which are the basic primitive forming the index sets or background space of medically imaged human anatomy. The geodesic flows of the submanifolds such as the landmarks, surface and subvolumes and the distance as measured by the geodesic flows of such coordinates, form the basic measuring and transporting tools of diffeomorphometry.

What is so important about the RKHS norm defining the kinetic energy in the action principle is that the vector fields of the geodesic motions of the submanifolds are superpositions of Green's Kernel's. For landmarks the superposition is a sum of weight kernels weighted by the canonical momentum which determines the inner product, for surfaces it is a surface integral, and for dense volumes it is a volume integral.

At $t = 0$ the geodesic has vector field $v_0 = Kp_0$ determined by the conjugate momentum and the Green's kernel of the inertial operator defining the Eulerian momentum $K = A^{-1}$. The metric distance between coordinate systems connected via the geodesic determined by the induced distance between identity and group element:

$$d_{Diff_V}(id, \varphi) = \| Log_{id}(\varphi) \|_V = \| v_0 \|_V = \sqrt{2H(id, p_0)}$$

Landmark and surface submanifolds have Lagrange multiplier associated to a sum and surface integral, respectively; dense volumes an integral with respect to Lebesgue measure.

Landmark or Pointset Geodesics

For Landmarks, the Hamiltonian momentum is defined on the indices, $p(i), i = 1, \ldots, n$ with the inner product given by $(p_t | v_t \circ \phi_t) \doteq \sum_i p_t(i) \cdot v_t \circ \phi_t(x_i)$ and Hamiltonian

$$H(\phi_t, p_t) = \frac{1}{2} \sum_j \sum_i p_t(i) \cdot K(\phi_t(x_i), \phi_t(x_j)) p_t(j).$$ The dynamics take the forms

$$\begin{cases} v_t = \sum_i K(\cdot, \phi_t(x_i)) p_t(i), \\ \dot{p}_t(i) = -(Dv_t)^T_{|\phi_t(x_i)} p_t(i), i = 1, 2, ; n \end{cases}$$

with the metric between landmarks $d^2 = (p_0 | v_0) = \sum_i p_0(i) \cdot \sum_j K(x_i, x_j) p_0(j).$

Surface Geodesics

For surfaces, the Hamiltonian momentum is defined across the surface with the inner product

$$(p_t | v_t \circ \phi_t) \doteq \int_U p_t(u) \cdot v_t \circ \phi_t(m(u)) du, \text{ with}$$

$$H(\phi_t, p_t) = \frac{1}{2} \int_U \int_U p_t(u) \cdot K(\phi_t(m(u)), \phi_t(m(v))) p_t(v) du dv.. \text{ The dynamics}$$

$$\begin{cases} v_t = \int_U K(\cdot, \phi_t(m(u))) p_t(u) du, \\ \dot{p}_t(u) = -(Dv_t)^T_{|\phi_t(m(u))} p_t(u), u \in U \end{cases}$$

with the metric between surface coordinates

$$d^2 = (p_0 | v_0) = \int_U p_0(u) \cdot \int_U K(m(u), m(u')) p_0(u') du du'$$

Volume Geodesics

For volumes the Hamiltonian momentum is $(p_t | v_t \circ \phi_t) \doteq \int_{\mathbb{R}^3} p_t(x) \cdot v_t \circ \phi_t(x) dx$ with

$H(\phi_t, p_t) = \frac{1}{2} \int_{\mathbb{R}^3} \int_{\mathbb{R}^3} p_t(x) \cdot K(\phi_t(x), \phi_t(y)) p_t(y) dx dy$. The dynamics

$$\begin{cases} v_t = \int_X K(\cdot, \phi_t(x)) p_t(x) dx, \\ \dot{p}_t(x) = -(Dv_t)^T_{|\phi_t(x)} p_t(x), x \in \mathbb{R}^3 \end{cases}$$

with the metric between volumes $d^2 = (p_0 \mid v_0) = \int_{\mathbb{R}^3} p_0(x) \cdot \int_{\mathbb{R}^3} K(x, y) p_0(y) dy\ dx$.

Conservation Laws on Diffeomorphic Shape Momentum for Computational Anatomy

Given the least-action there is a natural definition of momentum associated to generalized coordinates; the quantity acting against velocity gives energy. The field has studied two forms, the momentum associated to the Eulerian vector field termed Eulerian diffeomorphic shape momentum, and the momentum associated to the initial coordinates or canonical coordinates termed canonical diffeomorphic shape momentum. Each has a conservation law.The conservation of momentum goes hand in hand with the EL-General. In Computational anatomy, Av is the Eulerian Momentum since when integrated against Eulerian velocity v gives energy density; operator A the generalized moment of inertia or inertial operator which acting on the Eulerian velocity gives momentum which is conserved along the geodesic:

Eulerian $\qquad \dfrac{d}{dt}(Av_t \mid ((D\phi_t)w) \circ \phi_t^{-1}) = 0$, $t \in [0,1]$.

(Euler-Conservation-Constant-Energy)

Canonical $\qquad \dfrac{d}{dt}(p_t \mid (D\phi_t)w) = 0$, $t \in [0,1]$ for all $w \in V$.

Conservation of Eulerian shape momentum was shown in and follows from EL-General; conservation of canonical momentum was shown in

The proof follow from defining $w_t = ((D\phi_t)w) \circ \phi_t^{-1}$, $\dfrac{d}{dt} w_t = (Dv_t)w_t - (Dw_t)v_t$ implying

$$\frac{d}{dt}(Av_t \mid ((D\phi_t)w) \circ \phi_t^{-1}) = (\frac{d}{dt} Av_t \mid ((D\phi_t)w) \circ \phi_t^{-1}) + (Av_t \mid \frac{d}{dt}((D\phi_t)w) \circ \phi_t^{-1})$$

$$= (\frac{d}{dt} Av_t \mid w_t) + (Av_t \mid (Dv_t)w_t - (Dw_t)v_t) = 0.$$

The proof on Canonical momentum is shown from $\dot{p}_t = -(Dv_t)^T_{|\phi_t} p_t$:

$$\frac{d}{dt}(p_t \mid (D\phi_t)w) = (\dot{p}_t \mid (D\phi_t)w) + (p_t \mid \frac{d}{dt}(D\phi_t)w)$$

$$= (\dot{p}_t \,|\, (D\phi_t)w) + (p_t \,|\, (Dv_t)_{|_{\phi_t}} (D\phi_t)w) = 0.$$

Geodesic Interpolation of Information between Coordinate Systems Via Variational Problems

Construction of diffeomorphic correspondences between shapes calculates the initial vector field coordinates $v \in V$ and associated weights on the Greens kernels p_0. These initial coordinates are determined by matching of shapes, called Large Deformation Diffeomorphic Metric Mapping (LDDMM). LDDMM has been solved for landmarks with and without correspondence and for dense image matchings. curves, surfaces, dense vector and tensor imagery, and varifolds removing orientation. LDDMM calculates geodesic flows of the EL-General onto target coordinates, adding to the action integral $\frac{1}{2}\int_0^1 (Av_t \,|\, v_t)dt$ an endpoint matching condition $E : \phi_1 \to R^+$ measuring the correspon dence of elements in the orbit under coordinate system transformation. Existence of solutions were examined for image matching. The solution of the variational problem satisfies the EL-General for $t \in [0,1)$ with boundary condition.

Matching Based on Minimizing Kinetic Energy Action With Endpoint Condition

$$\min_{\phi : v = \dot{\phi} \circ \phi^{-1}, \phi_0 = id} C(\phi) \doteq \frac{1}{2}\int_0^1 (Av_t \,|\, v_t)dt + E(\phi_1)$$

Euler Conservation	$\dfrac{d}{dt} Av_t + ad_{v_t}^*(Av_t) = 0, t \in [0,1),$
Boundary Condition	$\phi_0 = id, Av_1 = -\dfrac{\partial E(\phi)}{\partial \phi}\big|_{\phi = \phi_1}.$

Conservation from EL-General extends the B.C. at $t = 1$ to the rest of the path $t \in [0,1)$. The inexact matching problem with the endpoint matching term $E(\phi_1)$ has several alternative forms. One of the key ideas of the stationarity of the Hamiltonian along the geodesic solution is the integrated running cost reduces to initial cost at t=0, geodesics of the EL-General are determined by their initial condition v_0.

The running cost is reduced to the initial cost determined by $v_0 = Kp_0$ of Kernel-Surf.-Land.-Geodesics.

Matching Based on Geodesic Shooting

$$\min \quad C(v_0) \quad -(Av_0 \,|\, v_0) + E(\mathrm{Exp}_{id}(v_0) \cdot I_0);$$

$$\min_{p_0} C(p_0) = \frac{1}{2}(p_0 \mid Kp_0) + E(\text{Exp}_{id}(Kp_0) \cdot I_0)$$

The matching problem explicitly indexed to initial condition v_0 is called shooting, which can also be reparamerized via the conjugate momentum p_0.

Dense Image Matching in Computational Anatomy

Dense image matching has a long history now with the earliest efforts exploiting a small deformation framework. Large deformations began in the early 90's, with the first existence to solutions to the variational problem for flows of diffeomorphisms for dense image matching established in. Beg solved via one of the earliest LDDMM algorithms based on solving the variational matching with endpoint defined by the dense imagery with respect to the vector fields, taking variations with respect to the vector fields. Another solution for dense image matching reparameterizes the optimization problem in terms of the state $q_t \doteq I \circ \phi_t^{-1}, q_0 = I$ giving the solution in terms of the infinitesimal action defined by the advection equation.

LDDMM Dense Image Matching

For Beg's LDDMM, denote the Image $I(x), x \in X$ with group action $\phi \cdot I \doteq I \circ \phi^{-1}$. Viewing this as an optimal control problem, the state of the system is the diffeomorphic flow of coordinates $\phi_t, t \in [0,1]$, with the dynamics relating the control $v_t, t \in [0,1]$ to the state given by $\dot{\phi} = v \circ \phi$. The endpoint matching condition $E(\phi_1) \doteq \| I \circ \phi_1^{-1} - I' \|^2$

gives the variational problem

$$\min_{v:\dot{\phi}=v \circ \phi} C(v) \doteq \frac{1}{2}\int_0^1 (Av_t \mid v_t)dt + \frac{1}{2}\int_{\mathbb{R}^3} |I \circ \phi_1^{-1}(x) - I'(x)|^2 \, dx \quad \begin{array}{l}\textit{(Dense-Image-Match-}\\\textit{ing)}\end{array}$$

$$\left\{\begin{array}{ll}\text{Endpoint Condition:} & Av_1 = \mu_1 dx, \mu_1 = (I \circ \phi_1^{-1} - I')\nabla(I \circ \phi_1^{-1}), \\ \text{Conservation:} & Av_t = \mu_t dx, \mu_t = (D\phi_t^{-1})^T \mu_0 \circ \phi_t^{-1} \mid D\phi_t^{-1} \mid. \\ & \mu_0 = (I - I' \circ \phi_1)\nabla I \mid D\phi_1 \mid. \end{array}\right.$$

Beg's iterative LDDMM algorithm has fixed points which satisfy the necessary optimizer conditions. The iterative algorithm is given in Beg's LDDMM algorithm for dense image matching.

Hamiltonian LDDMM in the Reduced Advected State

Denote the Image $I(x), x \in X$, with state $q_t \doteq I \circ \phi_t^{-1}$ and the dynamics related state and control given by the advective term $\dot{q}_t = -\nabla q_t \cdot v_t$. The endpoint $E(q_1) \doteq \| q_1 - I' \|^2$ gives the variational problem

$$\min_{q:\dot{q}=v\circ q} C(v) \doteq \frac{1}{2}\int_0^1 (Av_t \mid v_t)dt + \frac{1}{2}\int_{\mathbb{R}^3} |q_1(x)-I'(x)|^2\, dx \quad (\textit{Dense-Image-Matching})$$

Viallard's iterative Hamiltonian LDDMM has fixed points which satisfy the necessary optimizer conditions.

Diffusion Tensor Image Matching in Computational Anatomy

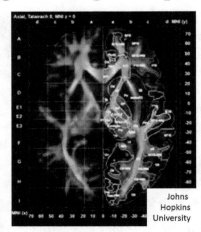

Image showing a diffusion tensor image with three color levels depicting the orientations of the three eigenvectors of the matrix image $I(x), x \in \mathbb{R}^2$, matrix valued image; each of three colors represents a direction.

Dense LDDMM tensor matching solves the variational problem matching between co-ordinate system based on the principle eigenvectors of the diffusion tensor MRI image (DTI) denoted $M(x), x \in \mathbb{R}^3$ consisting of the 3×3-tensor at every voxel. Several of the group actions defined based on the Frobenius matrix norm between square matrices $\|A\|_F^2 \doteq traceA^T A$. Shown in the accompanying figure is a DTI image illustrated via its color map depicting the eigenvector orientations of the DTI matrix at each voxel with color determined by the orientation of the directions.

Coordinate system transformation based on DTI imaging has exploited two actions, one based on the principle eigen-vector or entire matrix.

- LDDMM matching based on diffusion tensor matrix

Denote the 3×3 tensor image $M(x), x \in \mathbb{R}^3$ with eigen-elements $\{\lambda_i(x), e_i(x), i = 1, 2, 3\}$, eigenvalues $\lambda_1 \geq \lambda_2 \geq \lambda_3$, and e_1, e_2, e_3 eigenvectors. The group action becomes $\varphi \cdot M = (\lambda_1\hat{e}_1\hat{e}_1^T + \lambda_2\hat{e}_2\hat{e}_2^T + \lambda_3\hat{e}_3\hat{e}_3^T) \circ \varphi^{-1}$, transformed eigenvectors

$$\hat{e}_1 = \frac{D\varphi e_1}{\|D\varphi e_1\|}, \quad \hat{e}_2 = \frac{D\varphi e_2 - \langle \hat{e}_1, D\varphi e_2\rangle \hat{e}_1}{\sqrt{\|D\varphi e_2\|^2 - \langle \hat{e}_1, D\varphi e_2\rangle^2}}, \quad \hat{e}_3 = \hat{e}_1 \times \hat{e}_2,$$

with endpoint $E(\phi_1) \doteq \int_{\mathbb{R}^3} \| \phi_1 \cdot M(x) - M'(x) \|_F^2 \, dx$. The variational problem matching onto $M'(x), x \in \mathbb{R}$ becomes

$$\min_{v:v=\dot{\phi}\circ\phi^{-1}} \frac{1}{2}\int_0^1 (Av_t \mid v_t) dt + \alpha \int_{\mathbb{R}^3} \| \phi_1 \cdot M(x) - M'(x) \|_F^2 dx \qquad (\textit{Dense-TensorDTI-Matching})$$

- LDDMM matching based on the principal eigenvector of the diffusion tensor matrix

Denote the image $I(x), x \in \mathbb{R}^3$ taken as a unit vector field defined by the first eigenvector. The group action becomes

$$\varphi \cdot I = \begin{cases} \dfrac{D_{\varphi^{-1}}\varphi I \circ \varphi^{-1} \| I \circ \varphi^{-1} \|}{\| D_{\varphi^{-1}}\varphi I \circ \varphi^{-1} \|} & I \circ \varphi \neq 0; \\ 0 & \text{otherwise.} \end{cases}$$

with endpoint

$$E(\phi_1) \doteq \alpha \int_{\mathbb{R}^3} \| \phi_1 \cdot I - I' \|^2 dx + \beta \int_{\mathbb{R}^3} (\| \phi_1 \cdot I \| - \| I' \|)^2 dx).$$

The variational problem matching onto vector image $I'(x), x \in \mathbb{R}^3$ becomes

$$\min_{v:\dot{\phi}\circ\phi^{-1}} \frac{1}{2}\int_0^1 (Av_t \mid v_t) dt + \alpha \int_{\mathbb{R}^3} \| \phi_1 \cdot I - I' \|^2 dx + \beta \int_{\mathbb{R}^3} (\| \phi_1 \cdot I \| - \| I' \|)^2 dx.$$

Metamorphosis

Demonstrating metamorphosis allowing both diffeomorphic change in coordinate transformation as well as change in image intensity as associated to early Morphing technologies such as the Michael Jackson video. Notice the insertion of tumor gray level intensity which does not exist in template.

The principle mode of variation represented by the orbit model is change of coordinates. For setting in which pairs of images are not related by diffeomorphisms but have photometric variation or image variation not represented by the template, active appearance modelling has been introduced, originally by Edwards-Cootes-Taylor and in 3D medical imaging in. In the context of Computational Anatomy in which metrics on the anatomical orbit has been studied, metamorphosis for modelling structures such as tumors and photometric changes which are not resident in the template was introduced in for Magnetic Resonance image models, with many subsequent developments extending the metamorphosis framework.

For image matching the image metamorphosis framework enlarges the action so that $t \mapsto (\phi_t, I_t)$ with action $\phi_t \cdot I_t \doteq I_t \circ \phi_t^{-1}$. In this setting metamorphosis combines both the diffeomorphic coordinate system transformation of Computational Anatomy as well as the early morphing technologies which only faded or modified the photometric or image intensity alone.

Then the matching problem takes a form with equality boundary conditions:

$$\min_{(v,I)} \frac{1}{2} \int_0^1 \left((Av_t \mid v_t) + \| \dot{I}_t \circ \phi_t^{-1} \|^2 / \sigma^2 \right) dt \text{ subject to } \phi_0 = id, I_0 = \text{fixed}, I_1 = \text{fixed}$$

Matching Landmarks, Curves, Surfaces

Transforming coordinate systems based on Landmark point or fiducial marker features dates back to Bookstein's early work on small deformation spline methods for interpolating correspondences defined by fiducial points to the two-dimensional or three-dimensional background space in which the fiducials are defined. Large deformation landmark methods came on in the late 90's. The above Figure depicts a series of landmarks associated three brain structures, the amygdala, entorhinal cortex, and hippocampus.

Matching geometrical objects like unlabelled point distributions, curves or surfaces is another common problem in Computational Anatomy. Even in the discrete setting where these are commonly given as vertices with meshes, there are no predetermined correspondences between points as opposed to the situation of landmarks described above. From the theoretical point of view, while any submanifold X in $\mathbb{R}^3$, $d = 1, 2, 3$ can be parameterized in local charts $m : u \in U \subset \mathbb{R}^{0,1,2,3} \to \mathbb{R}^3$, all reparametrizations of these charts give geometrically the same manifold. Therefore, early on in Computational anatomy, investigators have identified the necessity of parametrization invariant representations. One indispensable requirement is that the end-point matching term between two submanifolds is itself independent of their parametrizations. This can be achieved via concepts and methods borrowed from Geometric measure theory, in particular currents and varifolds which have been used extensively for curve and surface matching.

Landmark or Point Matching with Correspondence

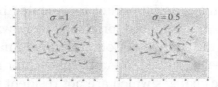

Figure showing landmark matching with correspondence. Left and right panels depict two different kernel with solutions.

Denoted the landmarked shape $X \doteq \{x_1,\ldots,x_n\} \subset \mathbb{R}^3$ with endpoint $E(\phi_1) \doteq \sum_i \|\phi_1(x_i) - x_i'\|^2$, the variational problem becomes

$$\min_{\phi:v=\dot{\phi}\circ\phi^{-1}} C(\phi) \doteq \frac{1}{2}\int (Av_t \mid v_t)dt + \frac{1}{2}\sum_i \|\phi_1(x_i) - x_i'\|^2 \qquad \textit{(Landmark-Matching)}$$

The geodesic Eulerian momentum is a generalized function $Av_t \in V^*, t \in [0,1]$, supported on the landmarked set in the variational problem. The endpoint condition with conservation implies the initial momentum at the identiy of the group:

Endpoint Condition: $Av_1 = \sum_{i=1}^{n} p_1(i)\delta_{\phi_1(x_i)}, p_1(i) = (x_i' - \phi_1(x_i)),$

Conservation: $Av_t = \sum_{i=1}^{n} p_t(i)\delta_{\phi_t(x_i)}, p_t(i) = (D\phi_{t1})^T_{|\phi_t(x_i)} p_1(i), \phi_{t1} \doteq \phi_1 \circ \phi_t^{-1},$

$Av_0 = \sum_i \delta_{x_i}(\cdot)p_0(i)$ with $p_0(i) = (D\phi_1)^T_{|x_i}(x_i' - \phi_1(x_i))$

The iterative algorithm for large deformation diffeomorphic metric mapping for landmarks is given.

Measure Matching: Unregistered Landmarks

Glaunes and co-workers first introduced diffeomorphic matching of pointsets in the general setting of matching distributions.

Curve Matching

In the one dimensional case, a curve in 3D can be represented by an embedding $m: u \in [0,1] \to \mathbb{R}^3$, and the group action of $Diff$ becomes $\phi \cdot m = \phi \circ m$. However, the correspondence between curves and embeddings is not one to one as the any reparametrization $m \circ \gamma$, for γ a diffeomorphism of the interval [0,1], represents geometrically the same curve. In order to preserve this invariance in the end-point matching term, several extensions of the previous 0-dimensional measure matching approach can be considered.

- Curve matching with currents

In the situation of oriented curves, currents give an efficient setting to construct invariant matching terms. In such representation, curves are interpreted as elements of a functional space dual to the space vector fields, and compared through kernel norms on these spaces. Matching of two curves m and m' writes eventually as the variational problem

$$\min_{\phi:v=\dot{\phi}\circ\phi^{-1}} C(\phi) \doteq \frac{1}{2}\int (Av_t \mid v_t)dt + \frac{1}{2}\|C_{\phi \cdot m} - C_{m'}\|^2_{cur}$$

with the endpoint term $E(\phi_1) = \| C_{\phi_1 \cdot m} - C_{m'} \|^2_{\text{cur}} / 2$ is obtained from the norm

$$\| C_m \|^2_{\text{cur}} = \int_0^1 \int_0^1 K_C(m(u), m(v)) \partial m(u) \cdot \partial m(v) du dv$$

the derivative $\partial m(u)$ being the tangent vector to the curve and K_C a given matrix kernel of $\mathbb{R}^3$. Such expressions are invariant to any positive reparametrizations of m and m', and thus still depend on the orientation of the two curves.

- Curve matching with varifolds

Varifold is an alternative to currents when orientation becomes an issue as for instance in situations involving multiple bundles of curves for which no "consistent" orientation may be defined. Varifolds directly extend 0-dimensional measures by adding an extra tangent space direction to the position of points, leading to represent curves as measures on the product of $\mathbb{R}^3$ and the Grassmannian of all straight lines in $\mathbb{R}^3$. The matching problem between two curves then consists in replacing the endpoint matching term by $E(\phi_1) = \| \mathcal{V}_{\phi_1 \cdot m} - \mathcal{V}_{m'} \|^2_{\text{cur}} / 2$ with varifold norms of the form:

$$\| \mathcal{V}_m \|^2_{var} = \int_0^1 \int_0^1 k_{\mathbb{R}^3}(m(u), m(v)) k_{\mathbf{Gr}}([\partial m(u)], [\partial m(v)]) \, |\partial m(u)| \, |\partial m(v)| du dv$$

where $[\partial m(u)]$ is the non-oriented line directed by tangent vector $\partial m(u)$ and $k_{\mathbb{R}^3}, k_{\mathbf{Gr}}$ two scalar kernels respectively on $\mathbb{R}^3$ and the Grassmannian. Due to the inherent non-oriented nature of the Grassmannian representation, such expressions are invariant to positive and negative reparametrizations.

Surface Matching

Surface matching share many similarities with the case of curves. Surfaces in $\mathbb{R}^3$ are parametrized in local charts by embeddings $m : u \in U \subset \mathbb{R}^2 \rightarrow \mathbb{R}^3$, with all reparametrizations $m \circ \gamma$ with γ a diffeomorphism of U being equivalent geometrically. Currents and varifolds can be also used to formalize surface matching.

- Surface matching with currents

Oriented surfaces can be represented as 2-currents which are dual to differential 2-forms. In $\mathbb{R}^3$, one can further identify 2-forms with vector fields through the standard wedge product of 3D vectors. In that setting, surface matching writes again:

$$\min_{\phi : v = \dot{\phi} \circ \phi^{-1}} C(\phi) \doteq \frac{1}{2} \int (A v_t \mid v_t) dt + \frac{1}{2} \| C_{\phi_1 \cdot m} - C_{m'} \|^2_{\text{cur}}$$

with the endpoint term $E(\phi_1) = \| C_{\phi_1 \cdot m} - C_{m'} \|^2_{\text{cur}} / 2$ given through the norm

$$\| C_m \|_{\text{cur}}^2 = \iint\limits_{U \times U} K_C(m(u), m(v)) \vec{n}(u) \cdot \vec{n}(v) du \, dv$$

with $\vec{n} = \partial_{u_1} m \wedge \partial_{u_2} m$ the normal vector to the surface parametrized by m.

- Surface matching with varifolds

For non-orientable or non-oriented surfaces, the varifold framework is often more adequate. Identifying the parametric surface m with a varifold V_m in the space of measures on the product of $\mathbb{R}^3$ and the Grassmannian, one simply replaces the previous current metric $\| C_m \|_{\text{cur}}^2$ by:

$$\| V_m \|_{\text{var}}^2 = \iint\limits_{U \times U} k_{\mathbb{R}^3}(m(u), m(v)) k_{\text{Gr}}([\vec{n}(u)], [\vec{n}(v)]) | \vec{n}(u) \| \vec{n}(v) | du \, dv$$

where $[\vec{n}(u)]$ is the (non-oriented) line directed by the normal vector to the surface.

Growth and Atrophy from Longitudinal Time-series

There are many settings in which there are a series of measurements, a time-series to which the underlying coordinate systems will be matched and flowed onto. This occurs for example in the dynamic growth and atrophy models and motion tracking such as have been explored in An observed time sequence is given and the goal is to infer the time flow of geometric change of coordinates carrying the exemplars or templars through the period of observations.

The generic time-series matching problem considers the series of times is $0 < t_1 < \ldots t_K = 1$. The flow optimizes at the series of costs $E(t_k), k = 1, \ldots, K$ giving optimization problems of the form

$$\min_{\phi : v = \dot{\phi} \circ \phi^{-1}, \phi_0 = id} C(\phi) \doteq \frac{1}{2} \int_0^1 (Av_t | v_t) dt + \sum_{k=1}^{K} E(\phi_{t_k}).$$

There have been at least three solutions offered thus far, piecewise geodesic, principal geodesic and splines.

The Random Orbit Model of Computational Anatomy

The random orbit model of Computational Anatomy first appeared in modelling the change in coordinates associated to the randomness of the group acting on the templates, which induces the randomness on the source of images in the anatomical orbit of shapes and forms and resulting observations through the medical imaging devices. Such a random orbit model in which randomness on the group induces randomness on the images was examined for the Special Euclidean Group for object recognition in.

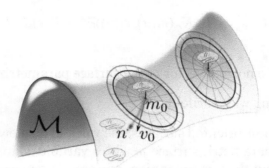

Orbits of brains associated to diffeomorphic group action on templates depicted via smooth flow associated to geodesic flows with random spray associatd to random generation of initial tangent space vector field $v_0 \in V$; published in.

Depicted in the figure is a depiction of the random orbits around each exemplar, $m_0 \in \mathcal{M}$, generated by randomizing the flow by generating the initial tangent space vector field at the identity $v_0 \in V$, and then generating random object $n \doteq Exp_{id}(v_0) \cdot m_0 \in \mathcal{M}$.

The random orbit model induces the prior on shapes and images $I \in \mathcal{I}$ conditioned on a particular atlas $I_a \in \mathcal{I}$. For this the generative model generates the mean field I as a random change in coordinates of the template according to $I \doteq \phi \cdot I_a$, where the diffeomorphic change in coordinates is generated randomly via the geodesic flows. The prior on random transformations $\pi_{Diff}(d\phi)$ on $Diff_V$ is induced by the flow $Exp_{id}(v)$, with $v \in V$ constructed as a Gaussian random field prior $\pi_V(dv)$. The density on the random observables at the output of the sensor $I^D \in \mathcal{I}^D$ are given by

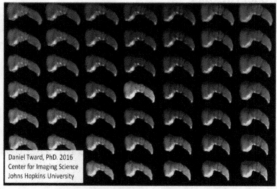

Figure showing the random spray of synthesized subcortical structures laid out in the two-dimensional grid representing the variance of the eigenfunction used for the momentum for synthesis.

$$p(I^D \mid I_a) = \int_V p(I^D \mid Exp_{id}(v) \cdot I_a)\pi_V(dv).$$

Shown in the Figure on the right the cartoon orbit, are a random spray of the subcortical manifolds generated by randomizing the vector fields v_0 supported over the submanifolds.

The Bayesian Model of Computational Anatomy

Source-channel model of CA

Anatomical Model		Distortion Process	
$I = \varphi \cdot I_{temp} \in \mathcal{I}$	$I \in \mathcal{I}$	(MRI, CT) $I^D = I + noise$	$I^D \in \mathcal{I}^D$

Source-channel model showing the source of images the deformable template $I \doteq \phi \cdot I_{temp} \in \mathcal{I}$ and channel output associated with MRI sensor $I^D \in \mathcal{I}^D$

The central statistical model of Computational Anatomy in the context of medical imaging has been the source-channel model of Shannon theory; the source is the deformable template of images $I \in \mathcal{I}$, the channel outputs are the imaging sensors with observables $I^D \in \mathcal{I}^D$ (see Figure). The importance of the source-channel model is that the variation in the anatomical configuration are modelled separated from the sensor variations of the Medical imagery. The Bayes theory dictates that the model is characterized by the prior on the source, $\pi_{\mathcal{I}}(\cdot)$ on $I \in \mathcal{I}$, and the conditional density on the observable $p(\cdot \mid I)$ on $I^D \in \mathcal{I}^D$ conditioned on $I \in \mathcal{I}$.

For image action $I(g) \doteq g \cdot I_{temp}, g \in \mathcal{G}$, then the prior on the group $\pi_{\mathcal{G}}(\cdot)$ induces the prior on images $\pi_{\mathcal{I}}(\cdot)$, written as densities the log-posterior takes the form

$$\log p(I(g) \mid I^D) \simeq \log p(I^D \mid I(g)) + \log \pi_{\mathcal{G}}(g).$$

MAP Estimation in the Multiple-atlas Orbit Model

Maximum a posteriori estimation (MAP) estimation is central to modern statistical theory. Parameters of interest $\theta \in \grave{E}$ take many forms including (i) disease type such as neurodegenerative or neurodevelopmental diseases, (ii) structure type such as cortical or subcorical structures in problems associated to segmentation of images, and (iii) template reconstruction from populations. Given the observed image I^D, MAP estimation maximizes the posterior:

$$\hat{\theta} \doteq \arg\max_{\theta \in \Theta} \log p(\theta \mid I^D).$$

This requires computation of the conditional probabilities $p(\theta \mid I^D) = \dfrac{p(I^D, \theta)}{p(I^D)}$. The

multiple atlas orbit model randomizes over the denumerable set of atlases $\{I_a, a \in \mathcal{A}\}$. The model on images in the orbit take the form of a multi-modal mixture distribution

$$p(I^D, \theta) = \sum_{a \in \mathcal{A}} p(I^D, \theta \mid I_a) \pi_A(a).$$

The conditional Gaussian model has been examined heavily for inexact matching in dense images and for alndmark matching.

- Dense Image Matching: Model $I^D(x), x \in X$ as a conditionally Gaussian random field conditioned, mean field, $\phi_1 \cdot I \doteq I(\phi_1^{-1}), \phi_1 \in Diff_V$. For uniform variance the endpoint error terms plays the role of the log-conditional (only a function of the mean field) giving the endpoint term:

$$-\log p(I^D \mid I(g)) \simeq E(\phi_1) \doteq \frac{1}{2\sigma^2} \| I^D - I \circ \phi_1^{-1} \|^2 . \qquad (Conditional\text{-}Gaussian)$$

- Landmark Matching: Model $Y = \{y_1, y_2, \ldots\}$ as conditionally Gaussian with mean field $\phi_1(x_i), i = 1, 2, \ldots, \phi_1 \in Diff_V$, constant noise variance independent of landmarks. The log-conditional (only a function of the mean field) can be viewed as the endpoint term:

$$-\log p(I^D \mid I(g)) \simeq E(\phi_1) \doteq \frac{1}{2\sigma^2} \sum_i \| y_i - \phi_1(x_i) \|^2.$$

MAP Segmentation Based on Multiple Atlases

The random orbit model for multiple atlases models the orbit of shapes as the union over multiple anatomical orbits generated from the group action of diffeomorphisms, $\mathcal{I} = \bigcup_{a \in A} Diff_V \cdot I_a$, with each atlas having a template and predefined segmentation field $(I_a, W_a), a = a_1, a_2, \ldots$ incorporating the parcellation into anatomical structures of the coordinate of the MRI.. The pairs are indexed over the voxel lattice $I_a(x_i), W_a(x_i), x_i \in X \subset \mathbb{R}^3$ with an MRI image and a dense labelling of every voxel coordinate. The anatomical labelling of parcellated structures are manual delineations by neuroanatomists.

The Bayes segmentation problem is given measurement I^D with mean field and parcellation (I, W), the anatomical labelling $\theta \doteq W$. mustg be estimated for the measured MRI image. The mean-field of the observable I^D image is modelled as a random deformation from one of the templates $I \doteq \varphi \cdot I_a$, which is also randomly selected, $A = a$,. The optimal diffeomorphism $\varphi \in \mathcal{G}$ is hidden and acts on the background space of coordinates of the randomly selected template image   . Given a single atlas a, the likelihood model for inference is determined by the joint probability $p(I^D, W \mid A = a)$; with multiple atlases, the fusion of the likelihood functions yields the multi-modal mixture model with the prior averaging over models.

The MAP estimator of segmentation W_a is the maximizer $\max_W \log p(W \mid I^D)$ given I^D, which involves the mixture over all atlases.

$$\hat{W} \doteq \arg\max_W \log p(I^D, W) \text{ with } p(I^D, W) = \sum_{a \in A} p(I^D, W \mid A = a)\pi_A(a).$$

The quantity $p(I^D, W)$ is computed via a fusion of likelihoods from multiple deformable atlases, with $\pi_A(a)$ being the prior probability that the observed image evolves from the specific template image I_a.

The MAP segmentation can be iteratively solved via the expectation-maximization(EM) algorithm

$$W^{\text{new}} \doteq \arg\max_W \int \log p(W, I^D, A, \varphi) dp(A, \varphi \mid W^{\text{old}}, I^D).$$

Statistical Shape theory in Computational Anatomy

Shape in computational anatomy is a local theory, indexing shapes and structures to templates to which they are bijectively mapped. Statistical shape in Computational Anatomy is the empirical study of diffeomorphic correspondences between populations and common template coordinate systems. Interestingly, this is a strong departure from Procrustes Analyses and shape theories pioneered by David G. Kendall in that the central group of Kendall's theories are the finite-dimensional Lie groups, whereas the theories of shape in Computational Anatomy have focussed on the diffeomorphism group, which to first order via the Jacobian can be thought of as a field - thus infinite dimensional - of low-dimensional Lie groups of scale and rotations.

Figure showing 100's of subcortical structures embedded in two-dimensional momentum space generated from the first two-eigenvectors of the empirical covariance estimated from the population of shapes.

The random orbit model provides the natural setting to understand empirical shape and shape statistics within Computational anatomy since the non-linearity of the induced probability law on anatomical shapes and forms $m \in \mathcal{M}$ is induced via the reduction to the vector fields $v_0 \in V$ at the tangent space at the identity of the diffeomorphism group. The successive flow of the Euler equation induces the random space of shapes and forms $Exp_{id}(v_0) \cdot m \in \mathcal{M}$.

Performing empirical statistics on this tangent space at the identity is the natural way for inducing probability laws on the statistics of shape. Since both the vector fields and the Eulerian momentum Av_0 are in a Hilbert space the natural model is one of a Gaussian random field, so that given test function $w \in V$, then the inner-products with the test functions are Gaussian distributed with mean and covariance.

This is depicted in the accompanying figure where subcortical brain structures are depicted in a two-dimensional coordinate system based on inner products of their initial vector fields that generate them from the template is shown in a 2-dimensional span of the Hilbert space.

Template Estimation from Populations

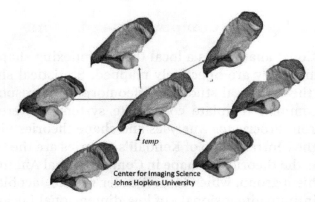

Center for Imaging Science
Johns Hopkins University

Depicting template estimation from multiplie subcortical surfaces in populations of MR images using the EM-algorithm solution of Ma.

The study of shape and statistics in populations are local theories, indexing shapes and structures to templates to which they are bijectively mapped. Statistical shape is then the study of diffeomorphic correspondences relative to the template. A core operation is the generation of templates from populations, estimating a shape that is matched to the population. There are several important methods for generating templates including methods based on Frechet averaging, and statistical approaches based on the expectation-maximization algorithm and the Bayes Random orbit models of Computational anatomy. Shown in the accompanying figure is a subcortical template reconstruction from the population of MRI subjects.

Software

Software suites containing a variety of diffeomorphic mapping algorithms include the following:

Cloud Software

- MRICloud

Neurocybernetics

In the physical sciences, neurocybernetics is the study of communication and automatic control systems in mutual relation to machines and living organisms. The underlying mathematical descriptions are control theory, extended for complex systems, and mean field theory for neural networks and neural field theory. Neurocybernetics is a sub-discipline of biocybernetics.

Etymology

Neurocybernetics is a compound word of *neuro*, the fundamental biological way to convey information within an organism by means of specially differentiated cells (neurons), and cybernetics, the science of communication and automatic control systems in relation to both machines and living beings.

Neuro-/biocybernetics can essentially be understood as the culmination of both neurology and cybernetics. As the complexity of neurology currently still prevents abstracting it into a generalizable theory, whilst on the other hand the complexity of cybernetical systems does not even come close to that of any biological system, even that of the most primitive kind (e.g. protozoa), neuro-/biocybernetics is still very much in the initial phase with much basic research going on, and hardly any commercial applications.

Generally speaking, it is the science that covers the integration of machines into a living organism via a Neural interface (aka neurolink or neural interface). The best example for applied neurocybernetics is the application of neuroprosthetics, which is still at a very early stage.

Introduction

The capacity of computers to cope with massive amounts of information and interface with each other with very low latencies is continually increasing. Efforts in the striving to advance human-computer interface technologies resulted in devices such as Virtual Reality gloves, various kind of motion trackers as well as 3-D sound and graphic based systems. These devices are capable of enhancing our ability to interact, along with novel approaches to user-interface-design, with vast amounts of information in as natural way as possible.

The emerging paradigm of human-machine interaction involves directly sensing bioelectric signals (from eye, muscle, the brain or any other nervous source) as inputs and rendering information in ways that take advantage of psycho-physiologic signal processing of the human nervous system (perceptual psychophysics).

After that the next step is to optimize the technology to the physiology, that is a biologically responsive interactive interface.

The Research

The ultimate goal of neurocybernetic research is the technological implementation of major principles of information processing in biological organisms by probing cellular and network mechanisms of brain functions. To unravel the biological design principles, computer-aided analyses of neuronal structure and signal transmission based on modern information theories and engineering methods are employed.

An offshoot of neurocybernetics is the field of *neurodynamics*, also called neural field theory, which uses differential equations to describe activity patterns in bulk neural matter. Research for neurodynamics involves the interdisciplinary areas of statistics and nonlinear physics and sensory neurobiology. On the physics side, topics of interest include information measures, oscillators, stochastic resonance, unstable periodic orbits, and pattern formation in ensembles of agents.

Practical Implementation

Practical applications, once the science has progressed, are countless but one especially remarkable would be neuroprosthetics that integrate seamlessly into the human organisms, by replicating and all layers of sensorial information from and to the surrogate organ. The demands of such a converter would be to preprocess the information and translate it via a synaptic bridge into information that is well adapted to the nervous system of the individual organism.

Some initial practical research is being undertaken. For example, in 2002, an array containing 100 electrodes, of which 25 could be accessed at any one time, was fired into the median nerve fibres of the scientist, Kevin Warwick. The neural signals obtained via the implant were detailed enough that a robotic arm developed by Warwick's colleague, Peter Kyberd, was able both to mimic the actions of Warwick's own arm and to provide a direct form of sensory feedback from fingertip sensors in the hand.

Other

Psycho-cybernetics is a self-help book written by plastic surgeon Maxwell Maltz and has nothing to do with neurocybernetics in the broader sense or any other science.

Neuroethology

Neuroethology is the evolutionary and comparative approach to the study of animal behavior and its underlying mechanistic control by the nervous system. This interdisciplinary branch of behavioral neuroscience endeavors to understand how the central nervous system translates biologically relevant stimuli into natural behavior. For example, many bats are capable of echolocation which is used for prey capture and navigation. The auditory system of bats is often cited as an example for how acoustic properties of sounds can be converted into a sensory map of behaviorally relevant features of sounds. Neuroethologists hope to uncover general principles of the nervous system from the study of animals with exaggerated or specialized behaviors.

Echolocation in bats is one model system in neuroethology

As its name implies, neuroethology is a multidisciplinary field composed of neurobiology (the study of the nervous system) and ethology (the study of behavior in natural conditions). A central theme of the field of neuroethology, delineating it from other branches of neuroscience, is this focus on natural behavior. Natural behaviors may be thought of as those behaviors generated through means of natural selection (i.e. finding mates, navigation, locomotion, predator avoidance) rather than behaviors in disease states, or behavioral tasks that are particular to the laboratory.

Philosophy

Neuroethology is an integrative approach to the study of animal behavior that draws upon several disciplines. Its approach stems from the theory that animals' nervous systems have evolved to address problems of sensing and acting in certain environmental niches and that their nervous systems are best understood in the context of the problems they have evolved to solve. In accordance with Krogh's principle, neuroethologists often study animals that are "specialists" in the behavior the researcher wishes to study e.g. honeybees and social behavior, bat echolocation, owl sound localization, etc.

The scope of neuroethological inquiry might be summarized by Jörg-Peter Ewert, a pioneer of neuroethology, when he considers the types of questions central to neuroethology in his 1980 introductory text to the field:

1. How are stimuli detected by an organism?

2. How are environmental stimuli in the external world represented in the nervous system?

3. How is information about a stimulus acquired, stored and recalled by the nervous system?

4. How is a behavioral pattern encoded by neural networks?

5. How is behavior coordinated and controlled by the nervous system?

6. How can the ontogenetic development of behavior be related to neural mechanisms?

Often central to addressing questions in neuroethology are comparative methodologies, drawing upon knowledge about related organisms' nervous systems, anatomies, life histories, behaviors and environmental niches. While it is not unusual for many types of neurobiology experiments to give rise to behavioral questions, many neuroethologists often begin their research programs by observing a species' behavior in its natural environment. Other approaches to understanding nervous systems include the systems identification approach, popular in engineering. The idea is to stimulate the system using a non-natural stimulus with certain properties. The system's response to the stimulus may be used to analyze the operation of the system. Such an approach is useful for linear systems, but the nervous system is notoriously nonlinear, and neuroethologists argue that such an approach is limited. This argument is supported by experiments in the auditory system. These experiments show that neural responses to complex sounds, like social calls, can not be predicted by the knowledge gained from studying the responses due to pure tones (one of the non-natural stimuli favored by auditory neurophysiologists). This is because of the non-linearity of the system.

Modern neuroethology is largely influenced by the research techniques used. Neural approaches are necessarily very diverse, as is evident through the variety of questions asked, measuring techniques used, relationships explored, and model systems employed. Techniques utilized since 1984 include the use of intracellular dyes, which make maps of identified neurons possible, and the use of brain slices, which bring vertebrate brains into better observation through intracellular electrodes (Hoyle 1984). Currently, other fields toward which neuroethology may be headed include computational neuroscience, molecular genetics, neuroendocrinology and epigenetics. The existing field of neural modeling may also expand into neuroethological terrain, due to its practical uses in robotics. In all this, neuroethologists must use the right level of simplicity to effectively guide research towards accomplishing the goals of neuroethology.

Critics of neuroethology might consider it a branch of neuroscience concerned with 'animal trivia'. Though neuroethological subjects tend not to be traditional neurobiological model systems (i.e. *Drosophila*, *C. elegans*, or *Danio rerio*), neuroethological approaches emphasizing comparative methods have uncovered many concepts central to neuroscience as a whole, such as lateral inhibition, coincidence detection, and sensory maps. The discipline of neuroethology has also discovered and explained the only vertebrate behavior for which the entire neural circuit has been described: the electric fish jamming avoidance response. Beyond its conceptual contributions, neuroethology makes indirect contributions to advancing human health. By understanding simpler nervous systems, many clinicians have used concepts uncovered by neuroethology and other branches of neuroscience to develop treatments for devastating human diseases.

History

The field of neuroethology owes part of its existence to the establishment of ethology as a unique discipline within the discipline of Zoology. Although animal behavior had been studied since the time of Aristotle (384-342 BC), it was not until the early twentieth century that ethology finally became distinguished from natural science (a strictly descriptive field) and ecology. The main catalysts behind this new distinction were the research and writings of Konrad Lorenz and Niko Tinbergen.

Konrad Lorenz was born in Austria in 1903, and is widely known for his contribution of the theory of fixed action patterns (FAPs): endogenous, instinctive behaviors involving a complex sequence of movements that are triggered ("released") by a certain kind of stimulus. This sequence always proceeds to completion, even if the original stimulus is removed. It is also species-specific and performed by nearly all members. Lorenz constructed his famous "hydraulic model" to help illustrate this concept, as well as the concept of action specific energy, or drives.

Niko Tinbergen was born in the Netherlands in 1907 and worked closely with Lorenz in the development of the FAP theory; their studies focused on the egg retrieval response of nesting geese. Tinbergen performed extensive research on the releasing mechanisms of particular FAPs, and used the bill-pecking behavior of baby herring gulls as his model system. This led to the concept of the supernormal stimulus. Tinbergen is also well known for his four questions that he believed ethologists should be asking about any given animal behavior; among these is that of the mechanism of the behavior, on a physiological, neural and molecular level, and this question can be thought of in many regards as the keystone question in neuroethology. Tinbergen also emphasized the need for ethologists and neurophysiologists to work together in their studies, a unity that has become a reality in the field of neuroethology.

Unlike behaviorism, which studied animals' reactions to non-natural stimuli in artificial, laboratory conditions, ethology sought to categorize and analyze the natural behaviors of animals in a field setting. Similarly, neuroethology asks questions about the neural bases of *naturally occurring* behaviors, and seeks to mimic the natural context as much as possible in the laboratory.

Although the development of ethology as a distinct discipline was crucial to the advent of neuroethology, equally important was the development of a more comprehensive understanding of Neuroscience. Contributors to this new understanding were the Spanish Neuroanatomist, Ramon y Cajal (born in 1852), and physiologists Charles Sherrington, Edgar Adrian, Alan Hodgkin, and Andrew Huxley. Charles Sherrington, who was born in Great Britain in 1857, is famous for his work on the nerve synapse as the site of transmission of nerve impulses, and for his work on reflexes in the spinal cord. His research also led him to hypothesize that every muscular activation is coupled to an inhibition of the opposing muscle. He was awarded a Nobel Prize for his work in 1932 along with Lord Edgar Adrian who made the first physiological recordings of neural activity from single nerve fibers.

Alan Hodgkin and Andrew Huxley (born 1914 and 1917, respectively, in Great Britain), are known for their collaborative effort to understand the production of action potentials in giant squid neurons. The pair also proposed the existence of ion channels to facilitate action potential initiation, and were awarded the Nobel Prize in 1963 for their efforts.

As a result of this pioneering research, many scientists then sought to connect the physiological aspects of the nervous and sensory systems to specific behaviors. These scientists – Karl von Frisch, Erich von Holst, and Theodore Bullock – are frequently referred to as the "fathers" of neuroethology. Neuroethology did not really come into its own, though, until the 1970s and 1980s, when new, sophisticated experimental methods allowed researchers such as Masakazu Konishi, Walter Heiligenberg, Jörg-Peter Ewert, and others to study the neural circuits underlying verifiable behavior.

Modern Neuroethology

The International Society for Neuroethology represents the present discipline of neuroethology, which was founded on the occasion of the NATO-Advanced Study Institute "Advances in Vertebrate Neuroethology" (August 13–24, 1981) organized by J.-P. Ewert, D.J. Ingle and R.R. Capranica, held at the University of Kassel in Hofgeismar, Germany (cf. report Trends in Neurosci. 5:141-143,1982). Its first president was Theodore H. Bullock. The society has met every three years since its first meeting in Tokyo in 1986.

Its membership draws from many research programs around the world; many of its members are students and faculty members from medical schools and neurobiology departments from various universities. Modern advances in neurophysiology techniques have enabled more exacting approaches in an ever-increasing number of animal systems, as size limitations are being dramatically overcome. Survey of the most recent (2007) congress of the ISN meeting symposia topics gives some idea of the field's breadth:

- Comparative aspects of spatial memory (rodents, birds, humans, bats)

- Influences of higher processing centers in active sensing (primates, owls, electric fish, rodents, frogs)

- Animal signaling plasticity over many time scales (electric fish, frogs, birds)

- Song production and learning in passerine birds

- Primate sociality

- Optimal function of sensory systems (flies, moths, frogs, fish)

- Neuronal complexity in behavior (insects, computational)

- Contributions of genes to behavior (Drosophila, honeybees, zebrafish)

- Eye and head movement (crustaceans, humans, robots)

- Hormonal actions in brain and behavior (rodents, primates, fish, frogs, and birds)

- Cognition in insects (honeybee)

Application to Technology

Neuroethology can help create advancements in technology through an advanced understanding of animal behavior. Model systems were generalized from the study of simple and related animals to humans. For example, the neuronal cortical space map discovered in bats, a specialized champion of hearing and navigating, elucidated the concept of a computational space map. In addition, the discovery of the space map in the barn owl led to the first neuronal example of the Jeffress model. This understanding is translatable to understanding spatial localization in humans, a mammalian relative of the bat. Today, knowledge learned from neuroethology are being applied in new technologies. For example, Randall Beer and his colleagues used algorithms learned from insect walking behavior to create robots designed to walk on uneven surfaces (Beer et al.). Neuroethology and technology contribute to one another bidirectionally.

Neuroethologists seek to understand the neural basis of a behavior as it would occur in an animal's natural environment but the techniques for neurophysiological analysis are lab-based, and cannot be performed in the field setting. This dichotomy between field and lab studies poses a challenge for neuroethology. From the neurophysiology perspective, experiments must be designed for controls and objective rigor, which contrasts with the ethology perspective—that the experiment be applicable to the animal's natural condition, which is uncontrolled, or subject to the dynamics of the environment. An early example of this is when Walter Rudolf Hess developed focal brain stimulation technique to examine a cat's brain controls of vegetative functions in addition to other behaviors. Even though this was a breakthrough in technological abilities and technique, it was not used by many neuroethologists originally because it compromised a cat's natural state, and, therefore, in their minds, devalued the experiments' relevance to real situations.

When intellectual obstacles like this were overcome, it led to a golden age of neuroethology, by focusing on simple and robust forms of behavior, and by applying modern neurobiological methods to explore the entire chain of sensory and neural mechanisms underlying these behaviors (Zupanc 2004). New technology allows neuroethologists to attach electrodes to even very sensitive parts of an animal such as its brain while it interacts with its environment. The founders of neuroethology ushered this understanding and incorporated technology and creative experimental design. Since then even indirect technological advancements such as battery-powered and waterproofed instruments have allowed neuroethologists to mimic natural conditions in the lab while they study behaviors objectively. In addition, the electronics required for amplifying neural signals and for transmitting them over a certain distance have enabled neuroscientists to record from behaving animals performing activities in naturalistic environments. Emerging technologies can complement neuroethology, augmenting the feasibility of this valuable perspective of natural neurophysiology.

Another challenge, and perhaps part of the beauty of neuroethology, is experimental design. The value of neuroethological criteria speak to the reliability of these experiments, because these discoveries represent behavior in the environments in which they evolved. Neuroethologists foresee future advancements through using new technologies and techniques, such as computational neuroscience, neuroendocrinology, and molecular genetics that mimic natural environments.

Case Studies

Jamming Avoidance Response

In 1963, two scientists, Akira Watanabe and Kimihisa Takeda, discovered the behavior of the jamming avoidance response in the knifefish *Eigenmannia* sp. In collaboration with T.H. Bullock and colleagues, the behavior was further developed. Finally, the work of W. Heiligenberg expanded it into a full neuroethology study by examining the series of neural connections that led to the behavior. *Eigenmannia* is a weakly electric fish that can self-generate electric discharges through electrocytes in its tail. Furthermore, it has the ability to electrolocate by analyzing the perturbations in its electric field. However, when the frequency of a neighboring fish's current is very close (less than 20 Hz difference) to that of its own, the fish will avoid having their signals interfere through a behavior known as Jamming Avoidance Response. If the neighbor's frequency is higher than the fish's discharge frequency, the fish will lower its frequency, and vice versa. The sign of the frequency difference is determined by analyzing the "beat" pattern of the incoming interference which consists of the combination of the two fish's discharge patterns.

Neuroethologists performed several experiments under *Eigenmannia*'s natural conditions to study how it determined the sign of the frequency difference. They manipulated the fish's discharge by injecting it with curare which prevented its natural electric organ from discharging. Then, an electrode was placed in its mouth and another was placed at the tip of its tail. Likewise, the neighboring fish's electric field was mimicked using another set of electrodes. This experiment allowed neuroethologists to manipulate different discharge frequencies and observe the fish's behavior. From the results, they were able to conclude that the electric field frequency, rather than an internal frequency measure, was used as a reference. This experiment is significant in that not only does it reveal a crucial neural mechanism underlying the behavior but also demonstrates the value neuroethologists place on studying animals in their natural habitats.

Feature Analysis in Toad Vision

The recognition of prey and predators in the toad was first studied in depth by Jörg-Peter Ewert. He began by observing the natural prey-catching behavior of the common toad (*Bufo bufo*) and concluded that the animal followed a sequence that consisted of stalking, binocular fixation, snapping, swallowing and mouth-wiping. However, initially, the toad's actions were dependent on specific features of the sensory stimulus: whether it demonstrated worm or anti-worm configurations. It was observed that the worm configuration, which signaled prey, was initiated by movement along the object's long axis, whereas anti-worm configuration, which signaled predator, was due to movement along the short axis. (Zupanc 2004).

Ewert and coworkers adopted a variety of methods to study the predator versus prey behavior response. They conducted recording experiments where they inserted electrodes into the brain, while the toad was presented with worm or anti-worm stimuli. This technique was repeated at different levels of the visual system and also allowed feature detectors to be identified. In focus was the discovery of prey-selective neurons in the optic tectum, whose axons could be traced towards the snapping pattern generating cells in the hypoglossal nucleus. The discharge patterns of prey-selective tectal neurons in response to prey objects – in freely moving toads – "predicted" prey-catching reactions such as snapping. Another approach, called stimulation experiment, was carried out in freely moving toads. Focal electrical stimuli were applied to different regions of the brain, and the toad's response was observed. When the thalamic-pretectal region was stimulated, the toad exhibited escape responses, but when the tectum was stimulated in an area close to prey-selective neurons, the toad engaged in prey catching behavior (Carew 2000). Furthermore, neuroanatomical experiments were carried out where the toad's thalamic-pretectal/tectal connection was lesioned and the resulting deficit noted: the prey-selective properties were abolished both in the responses of prey-selective neurons and in the prey catching behavior. These and other experiments suggest that prey selectivity results from pretecto-tectal influences.

Ewert and coworkers showed in toads that there are stimulus-response mediating pathways that translate perception (of visual sign stimuli) into action (adequate behavioral responses). In addition there are modulatory loops that initiate, modify or specify this mediation (Ewert 2004). Regarding the latter, for example, the telencephalic caudal ventral striatum is involved in a loop gating the stimulus-response mediation in a manner of directed attention. The telencephalic ventral medial pallium ("primordium hippocampi"), however, is involved in loops that either modify prey-selection due to associative learning or specify prey-selection due to non-associative learning, respectively.

Computational Neuroethology

Computational neuroethology (CN or CNE) is concerned with the computer modelling of the neural mechanisms underlying animal behaviors. Computational neuroethology was first argued for in depth by Randall Beer and by Dave Cliff both of whom acknowledged the strong influence of Michael Arbib's *Rana Computatrix* computational model of neural mechanisms for visual guidance in frogs and toads.

CNE systems work within a closed-loop environment; that is, they perceive their (perhaps artificial) environment directly, rather than through human input, as is typical in AI systems. For example, Barlow et al. developed a time-dependent model for the retina of the horseshoe crab *Limulus polyphemus* on a Connection Machine (Model CM-2). Instead of feeding the model retina with idealized input signals, they exposed the simulation to digitized video sequences made underwater, and compared its response with those of real animals.

Model Systems

- Bat echolocation – nocturnal flight navigation and prey capture; location of objects using echo returns of its own call

- Oscine bird song – zebra finch (*Taeniopygia guttata*), canary (*Serinus canaria*) and white-crowned sparrow (*Zonotrichia leucophrys*); song learning as a model for human speech development

- Electric fish – navigation, communication, Jamming Avoidance Response (JAR), corollary discharge, expectation generators, and spike timing dependent plasticity

- Barn owl auditory spatial map – nocturnal prey location and capture

- Toad vision – discrimination of prey versus predator – Video "Image processing in the toad's visual system: behavior, brain function, artificial neuronal net"

- Circadian rhythm – influence of various circadian controlled behaviors by the suprachiasmatic nucleus

- Cricket song – mate attraction and corollary discharge

- Fish Mauthner cells – C-start escape response and underwater directional hearing

- Fly – Microscale directional hearing in *Ormia ochracea*, sex differences of the visual system in Bibionidae, and spatial navigation in chasing behavior of Fannia canicularis

- Noctuid moths – ultrasound avoidance response to bat calls

- Aplysia – learning and memory in startle response

- Rat – spatial memory and navigation

- Salmon homing – olfactory imprinting and thyroid hormones

- Crayfish – escape and startle behaviors, aggression and formation of social hierarchies

- Cichlid fish – aggression and attack behaviors

- Honey bee – learning, navigation, vision, olfaction, flight, aggression, foraging

- Monarch butterfly – navigational mechanisms

- More Model Systems and Information

Neural Coding

Neural coding is a neuroscience related field concerned with characterizing the relationship between the stimulus and the individual or ensemble neuronal responses and the relationship among the electrical activity of the neurons in the ensemble. Based on the theory that sensory and other information is represented in the brain by networks of neurons, it is thought that neurons can encode both digital and analog information.

Overview

Neurons are remarkable among the cells of the body in their ability to propagate signals rapidly over large distances. They do this by generating characteristic electrical pulses called action potentials: voltage spikes that can travel down nerve fibers. Sensory neurons change their activities by firing sequences of action potentials in various temporal patterns, with the presence of external sensory stimuli, such as light, sound, taste, smell and touch. It is known that information about the stimulus is encoded in this pattern of action potentials and transmitted into and around the brain.

Although action potentials can vary somewhat in duration, amplitude and shape, they are typically treated as identical stereotyped events in neural coding studies. If the brief duration of an action potential (about 1ms) is ignored, an action potential sequence, or spike train, can be characterized simply by a series of all-or-none point events in time. The lengths of interspike intervals (ISIs) between two successive spikes in a spike train often vary, apparently randomly. The study of neural coding involves measuring and characterizing how stimulus attributes, such as light or sound intensity, or motor actions, such as the direction of an arm movement, are represented by neuron action potentials or spikes. In order to describe and analyze neuronal firing, statistical methods and methods of probability theory and stochastic point processes have been widely applied.

With the development of large-scale neural recording and decoding technologies, re-searchers have begun to crack the neural code and already provided the first glimpse into the real-time neural code as memory is formed and recalled in the hippocampus, a brain region known to be central for memory formation. Neuroscientists have initiated several large-scale brain decoding projects.

Encoding and Decoding

The link between stimulus and response can be studied from two opposite points of view. Neural encoding refers to the map from stimulus to response. The main focus is to understand how neurons respond to a wide variety of stimuli, and to construct models that attempt to predict responses to other stimuli. Neural decoding refers to the reverse map, from response to stimulus, and the challenge is to reconstruct a stimulus, or certain aspects of that stimulus, from the spike sequences it evokes.

Coding Schemes

A sequence, or 'train', of spikes may contain information based on different cod-ing schemes. In motor neurons, for example, the strength at which an innervated muscle is flexed depends solely on the 'firing rate', the average number of spikes per unit time (a 'rate code'). At the other end, a complex 'temporal code' is based on the precise timing of single spikes. They may be locked to an external stimulus such as in the visual and auditory system or be generated intrinsically by the neural circuitry.

Whether neurons use rate coding or temporal coding is a topic of intense debate within the neuroscience community, even though there is no clear definition of what these terms mean. In one theory, termed "neuroelectrodynamics", the following coding schemes are all considered to be epiphenomena, replaced instead by molecular chang-es reflecting the spatial distribution of electric fields within neurons as a result of the broad electromagnetic spectrum of action potentials, and manifested in information as spike directivity.

Rate Coding

The rate coding model of neuronal firing communication states that as the intensity of a stimulus increases, the frequency or rate of action potentials, or "spike firing", increas-es. Rate coding is sometimes called frequency coding.

Rate coding is a traditional coding scheme, assuming that most, if not all, information about the stimulus is contained in the firing rate of the neuron. Because the sequence of action potentials generated by a given stimulus varies from trial to trial, neuronal responses are typically treated statistically or probabilistically. They may be character-ized by firing rates, rather than as specific spike sequences. In most sensory systems, the firing rate increases, generally non-linearly, with increasing stimulus intensity.

Any information possibly encoded in the temporal structure of the spike train is ignored. Consequently, rate coding is inefficient but highly robust with respect to the ISI 'noise'.

During rate coding, precisely calculating firing rate is very important. In fact, the term "firing rate" has a few different definitions, which refer to different averaging procedures, such as an average over time or an average over several repetitions of experiment.

In rate coding, learning is based on activity-dependent synaptic weight modifications.

Rate coding was originally shown by ED Adrian and Y Zotterman in 1926. In this simple experiment different weights were hung from a muscle. As the weight of the stimulus increased, the number of spikes recorded from sensory nerves innervating the muscle also increased. From these original experiments, Adrian and Zotterman concluded that action potentials were unitary events, and that the frequency of events, and not individual event magnitude, was the basis for most inter-neuronal communication.

In the following decades, measurement of firing rates became a standard tool for describing the properties of all types of sensory or cortical neurons, partly due to the relative ease of measuring rates experimentally. However, this approach neglects all the information possibly contained in the exact timing of the spikes. During recent years, more and more experimental evidence has suggested that a straightforward firing rate concept based on temporal averaging may be too simplistic to describe brain activity.

Spike-count Rate

The Spike-count rate, also referred to as temporal average, is obtained by counting the number of spikes that appear during a trial and dividing by the duration of trial. The length T of the time window is set by experimenter and depends on the type of neuron recorded from and the stimulus. In practice, to get sensible averages, several spikes should occur within the time window. Typical values are T = 100 ms or T = 500 ms, but the duration may also be longer or shorter.

The spike-count rate can be determined from a single trial, but at the expense of losing all temporal resolution about variations in neural response during the course of the trial. Temporal averaging can work well in cases where the stimulus is constant or slowly varying and does not require a fast reaction of the organism — and this is the situation usually encountered in experimental protocols. Real-world input, however, is hardly stationary, but often changing on a fast time scale. For example, even when viewing a static image, humans perform saccades, rapid changes of the direction of gaze. The image projected onto the retinal photoreceptors changes therefore every few hundred milliseconds.

Despite its shortcomings, the concept of a spike-count rate code is widely used not only in experiments, but also in models of neural networks. It has led to the idea that a neuron transforms information about a single input variable (the stimulus strength) into a single continuous output variable (the firing rate).

There is a growing body of evidence that in Purkinje neurons, at least, information is not simply encoded in firing but also in the timing and duration of non-firing, quiescent periods.

Time-dependent Firing Rate

The time-dependent firing rate is defined as the average number of spikes (averaged over trials) appearing during a short interval between times t and t+Δt, divided by the duration of the interval. It works for stationary as well as for time-dependent stimuli. To experimentally measure the time-dependent firing rate, the experimenter records from a neuron while stimulating with some input sequence. The same stimulation sequence is repeated several times and the neuronal response is reported in a Peri-Stimulus-Time Histogram (PSTH). The time t is measured with respect to the start of the stimulation sequence. The Δt must be large enough (typically in the range of one or a few milliseconds) so there are sufficient number of spikes within the interval to obtain a reliable estimate of the average. The number of occurrences of spikes $n_K(t; t+\Delta t)$ summed over all repetitions of the experiment divided by the number K of repetitions is a measure of the typical activity of the neuron between time t and t+Δt. A further division by the interval length Δt yields time-dependent firing rate r(t) of the neuron, which is equivalent to the spike density of PSTH.

For sufficiently small Δt, r(t)Δt is the average number of spikes occurring between times t and t+Δt over multiple trials. If Δt is small, there will never be more than one spike within the interval between t and t+Δt on any given trial. This means that r(t)Δt is also the fraction of trials on which a spike occurred between those times. Equivalently, r(t)Δt is the probability that a spike occurs during this time interval.

As an experimental procedure, the time-dependent firing rate measure is a useful method to evaluate neuronal activity, in particular in the case of time-dependent stimuli. The obvious problem with this approach is that it can not be the coding scheme used by neurons in the brain. Neurons can not wait for the stimuli to repeatedly present in an exactly same manner before generating response.

Nevertheless, the experimental time-dependent firing rate measure can make sense, if there are large populations of independent neurons that receive the same stimulus. Instead of recording from a population of N neurons in a single run, it is experimentally easier to record from a single neuron and average over N repeated runs. Thus, the time-dependent firing rate coding relies on the implicit assumption that there are always populations of neurons.

Temporal Coding

When precise spike timing or high-frequency firing-rate fluctuations are found to carry information, the neural code is often identified as a temporal code. A number of studies have found that the temporal resolution of the neural code is on a millisecond time scale, indicating that precise spike timing is a significant element in neural coding.

Neurons exhibit high-frequency fluctuations of firing-rates which could be noise or could carry information. Rate coding models suggest that these irregularities are noise, while temporal coding models suggest that they encode information. If the nervous system only used rate codes to convey information, a more consistent, regular firing rate would have been evolutionarily advantageous, and neurons would have utilized this code over other less robust options. Temporal coding supplies an alternate explanation for the "noise," suggesting that it actually encodes information and affects neural processing. To model this idea, binary symbols can be used to mark the spikes: 1 for a spike, 0 for no spike. Temporal coding allows the sequence 000111000111 to mean something different from 001100110011, even though the mean firing rate is the same for both sequences, at 6 spikes/10 ms. Until recently, scientists had put the most emphasis on rate encoding as an explanation for post-synaptic potential patterns. However, functions of the brain are more temporally precise than the use of only rate encoding seems to allow. In other words, essential information could be lost due to the inability of the rate code to capture all the available information of the spike train. In addition, responses are different enough between similar (but not identical) stimuli to suggest that the distinct patterns of spikes contain a higher volume of information than is possible to include in a rate code.

Temporal codes employ those features of the spiking activity that cannot be described by the firing rate. For example, time to first spike after the stimulus onset, characteristics based on the second and higher statistical moments of the ISI probability distribution, spike randomness, or precisely timed groups of spikes (temporal patterns) are candidates for temporal codes. As there is no absolute time reference in the nervous system, the information is carried either in terms of the relative timing of spikes in a population of neurons or with respect to an ongoing brain oscillation.

The temporal structure of a spike train or firing rate evoked by a stimulus is determined both by the dynamics of the stimulus and by the nature of the neural encoding process. Stimuli that change rapidly tend to generate precisely timed spikes and rapidly changing firing rates no matter what neural coding strategy is being used. Temporal coding refers to temporal precision in the response that does not arise solely from the dynamics of the stimulus, but that nevertheless relates to properties of the stimulus. The interplay between stimulus and encoding dynamics makes the identification of a temporal code difficult.

In temporal coding, learning can be explained by activity-dependent synaptic delay modifications. The modifications can themselves depend not only on spike rates (rate coding) but also on spike timing patterns (temporal coding), i.e., can be a special case of spike-timing-dependent plasticity.

The issue of temporal coding is distinct and independent from the issue of independent-spike coding. If each spike is independent of all the other spikes in the train, the temporal character of the neural code is determined by the behavior of time-dependent firing rate r(t). If r(t) varies slowly with time, the code is typically called a rate code, and if it varies rapidly, the code is called temporal.

Temporal Coding in Sensory Systems

For very brief stimuli, a neuron's maximum firing rate may not be fast enough to produce more than a single spike. Due to the density of information about the abbreviated stimulus contained in this single spike, it would seem that the timing of the spike itself would have to convey more information than simply the average frequency of action potentials over a given period of time. This model is especially important for sound localization, which occurs within the brain on the order of milliseconds. The brain must obtain a large quantity of information based on a relatively short neural response. Additionally, if low firing rates on the order of ten spikes per second must be distinguished from arbitrarily close rate coding for different stimuli, then a neuron trying to discriminate these two stimuli may need to wait for a second or more to accumulate enough information. This is not consistent with numerous organisms which are able to discriminate between stimuli in the time frame of milliseconds, suggesting that a rate code is not the only model at work.

To account for the fast encoding of visual stimuli, it has been suggested that neurons of the retina encode visual information in the latency time between stimulus onset and first action potential, also called latency to first spike. This type of temporal coding has been shown also in the auditory and somato-sensory system. The main drawback of such a coding scheme is its sensitivity to intrinsic neuronal fluctuations. In the primary visual cortex of macaques, the timing of the first spike relative to the start of the stimulus was found to provide more information than the interval between spikes. However, the inter-spike interval could be used to encode additional information, which is especially important when the spike rate reaches its limit, as in high-contrast situations. For this reason, temporal coding may play a part in coding defined edges rather than gradual transitions.

The mammalian gustatory system is useful for studying temporal coding because of its fairly distinct stimuli and the easily discernible responses of the organism. Temporally encoded information may help an organism discriminate between different tastants of the same category (sweet, bitter, sour, salty, umami) that elicit very similar responses in terms of spike count. The temporal component of the pattern elicited by each tastant may be used to determine its identity (e.g., the difference between two bitter tastants, such as quinine and denatonium). In this way, both rate coding and temporal coding may be used in the gustatory system – rate for basic tastant type, temporal for more specific differentiation. Research on mammalian gustatory system has shown that there is an abundance of information present in temporal patterns across populations of neurons, and this information is different from that which is determined by rate coding schemes. Groups of neurons may synchronize in response to a stimulus. In studies dealing with the front cortical portion of the brain in primates, precise patterns with short time scales only a few milliseconds in length were found across small populations of neurons which correlated with certain information processing behaviors. However, little information could be determined from the patterns; one possible theory is they represented the higher-order processing taking place in the brain.

As with the visual system, in mitral/tufted cells in the olfactory bulb of mice, first-spike latency relative to the start of a sniffing action seemed to encode much of the information about an odor. This strategy of using spike latency allows for rapid identification of and reaction to an odorant. In addition, some mitral/tufted cells have specific firing patterns for given odorants. This type of extra information could help in recognizing a certain odor, but is not completely necessary, as average spike count over the course of the animal's sniffing was also a good identifier. Along the same lines, experiments done with the olfactory system of rabbits showed distinct patterns which correlated with different subsets of odorants, and a similar result was obtained in experiments with the locust olfactory system.

Temporal Coding Applications

The specificity of temporal coding requires highly refined technology to measure informative, reliable, experimental data. Advances made in optogenetics allow neurologists to control spikes in individual neurons, offering electrical and spatial single-cell resolution. For example, blue light causes the light-gated ion channel channelrhodopsin to open, depolarizing the cell and producing a spike. When blue light is not sensed by the cell, the channel closes, and the neuron ceases to spike. The pattern of the spikes matches the pattern of the blue light stimuli. By inserting channelrhodopsin gene sequences into mouse DNA, researchers can control spikes and therefore certain behaviors of the mouse (e.g., making the mouse turn left). Researchers, through optogenetics, have the tools to effect different temporal codes in a neuron while maintaining the same mean firing rate, and thereby can test whether or not temporal coding occurs in specific neural circuits.

Optogenetic technology also has the potential to enable the correction of spike abnormalities at the root of several neurological and psychological disorders. If neurons do encode information in individual spike timing patterns, key signals could be missed by attempting to crack the code while looking only at mean firing rates. Understanding any temporally encoded aspects of the neural code and replicating these sequences in neurons could allow for greater control and treatment of neurological disorders such as depression, schizophrenia, and Parkinson's disease. Regulation of spike intervals in single cells more precisely controls brain activity than the addition of pharmacological agents intravenously.

Phase-of-firing Code

Phase-of-firing code is a neural coding scheme that combines the spike count code with a time reference based on oscillations. This type of code takes into account a time label for each spike according to a time reference based on phase of local ongoing oscillations at low or high frequencies. A feature of this code is that neurons adhere to a preferred order of spiking, resulting in firing sequence.

It has been shown that neurons in some cortical sensory areas encode rich naturalistic stimuli in terms of their spike times relative to the phase of ongoing network fluctuations, rather than only in terms of their spike count. Oscillations reflect local field potential signals. It is often categorized as a temporal code although the time label used for spikes is coarse grained. That is, four discrete values for phase are enough to represent all the information content in this kind of code with respect to the phase of oscillations in low frequencies. Phase-of-firing code is loosely based on the phase precession phenomena observed in place cells of the hippocampus.

Phase code has been shown in visual cortex to involve also high-frequency oscillations. Within a cycle of gamma oscillation, each neuron has its own preferred relative firing time. As a result, an entire population of neurons generates a firing sequence that has a duration of up to about 15 ms.

Population Coding

Population coding is a method to represent stimuli by using the joint activities of a number of neurons. In population coding, each neuron has a distribution of responses over some set of inputs, and the responses of many neurons may be combined to determine some value about the inputs.

From the theoretical point of view, population coding is one of a few mathematically well-formulated problems in neuroscience. It grasps the essential features of neural coding and yet is simple enough for theoretic analysis. Experimental studies have revealed that this coding paradigm is widely used in the sensor and motor areas of the brain. For example, in the visual area medial temporal (MT), neurons are tuned to the moving direction. In response to an object moving in a particular direction, many neurons in MT fire with a noise-corrupted and bell-shaped activity pattern across the population. The moving direction of the object is retrieved from the population activity, to be immune from the fluctuation existing in a single neuron's signal. In one classic example in the primary motor cortex, Apostolos Georgopoulos and colleagues trained monkeys to move a joystick towards a lit target. They found that a single neuron would fire for multiple target directions. However it would fire fastest for one direction and more slowly depending on how close the target was to the neuron's 'preferred' direction.

Kenneth Johnson originally derived that if each neuron represents movement in its preferred direction, and the vector sum of all neurons is calculated (each neuron has a firing rate and a preferred direction), the sum points in the direction of motion. In this manner, the population of neurons codes the signal for the motion. This particular population code is referred to as population vector coding. This particular study divided the field of motor physiologists between Evarts' "upper motor neuron" group, which followed the hypothesis that motor cortex neurons contributed to control of single muscles, and the Georgopoulos group studying the representation of movement directions in cortex.

The Johns Hopkins University Neural Encoding laboratory led by Murray Sachs and Eric Young developed place-time population codes, termed the Averaged-Localized-Synchronized-Response (ALSR) code for neural representation of auditory acoustic stimuli. This exploits both the place or tuning within the auditory nerve, as well as the phase-locking within each nerve fiber Auditory nerve. The first ALSR reprsentation was for steady-state vowels; ALSR representations of pitch and formant frequencies in complex, non-steady state stimuli were demonstrated for voiced-pitch and formant representations in consonant-vowel syllables. The advantage of such representations is that global features such as pitch or formant transition profiles can be represented as global features across the entire nerve simultaneously via both rate and place coding.

Population coding has a number of other advantages as well, including reduction of uncertainty due to neuronal variability and the ability to represent a number of different stimulus attributes simultaneously. Population coding is also much faster than rate coding and can reflect changes in the stimulus conditions nearly instantaneously. Individual neurons in such a population typically have different but overlapping selectivities, so that many neurons, but not necessarily all, respond to a given stimulus.

Typically an encoding function has a peak value such that activity of the neuron is greatest if the perceptual value is close to the peak value, and becomes reduced accordingly for values less close to the peak value.

It follows that the actual perceived value can be reconstructed from the overall pattern of activity in the set of neurons. The Johnson/Georgopoulos vector coding is an example of simple averaging. A more sophisticated mathematical technique for performing such a reconstruction is the method of maximum likelihood based on a multivariate distribution of the neuronal responses. These models can assume independence, second order correlations , or even more detailed dependencies such as higher order maximum entropy models or copulas.

Correlation Coding

The correlation coding model of neuronal firing claims that correlations between action potentials, or "spikes", within a spike train may carry additional information above and beyond the simple timing of the spikes. Early work suggested that correlation between spike trains can only reduce, and never increase, the total mutual information present in the two spike trains about a stimulus feature. However, this was later demonstrated to be incorrect. Correlation structure can increase information content if noise and signal correlations are of opposite sign. Correlations can also carry information not present in the average firing rate of two pairs of neurons. A good example of this exists in the pentobarbital-anesthetized marmoset auditory cortex, in which a pure tone causes an increase in the number of correlated spikes, but not an increase in the mean firing rate, of pairs of neurons.

Independent-spike Coding

The independent-spike coding model of neuronal firing claims that each individual action potential, or "spike", is independent of each other spike within the spike train.

Position Coding

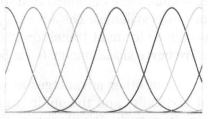

Plot of typical position coding

A typical population code involves neurons with a Gaussian tuning curve whose means vary linearly with the stimulus intensity, meaning that the neuron responds most strongly (in terms of spikes per second) to a stimulus near the mean. The actual intensity could be recovered as the stimulus level corresponding to the mean of the neuron with the greatest response. However, the noise inherent in neural responses means that a maximum likelihood estimation function is more accurate.

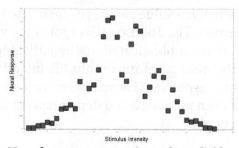

Neural responses are noisy and unreliable.

This type of code is used to encode continuous variables such as joint position, eye position, color, or sound frequency. Any individual neuron is too noisy to faithfully encode the variable using rate coding, but an entire population ensures greater fidelity and precision. For a population of unimodal tuning curves, i.e. with a single peak, the precision typically scales linearly with the number of neurons. Hence, for half the precision, half as many neurons are required. In contrast, when the tuning curves have multiple peaks, as in grid cells that represent space, the precision of the population can scale exponentially with the number of neurons. This greatly reduces the number of neurons required for the same precision.

Sparse Coding

The sparse code is when each item is encoded by the strong activation of a relatively small set of neurons. For each item to be encoded, this is a different subset of all available neurons.

As a consequence, sparseness may be focused on temporal sparseness ("a relatively small number of time periods are active") or on the sparseness in an activated population of neurons. In this latter case, this may be defined in one time period as the number of activated neurons relative to the total number of neurons in the population. This seems to be a hallmark of neural computations since compared to traditional computers, information is massively distributed across neurons. A major result in neural coding from Olshausen and Field is that sparse coding of natural images produces wavelet-like oriented filters that resemble the receptive fields of simple cells in the visual cortex. The capacity of sparse codes may be increased by simultaneous use of temporal coding, as found in the locust olfactory system.

Given a potentially large set of input patterns, sparse coding algorithms (e.g. Sparse Autoencoder) attempt to automatically find a small number of representative patterns which, when combined in the right proportions, reproduce the original input patterns. The sparse coding for the input then consists of those representative patterns. For example, the very large set of English sentences can be encoded by a small number of symbols (i.e. letters, numbers, punctuation, and spaces) combined in a particular order for a particular sentence, and so a sparse coding for English would be those symbols.

Linear Generative Model

The codings generated by algorithms implementing a linear generative model can be classified into codings with *soft sparseness* and those with *hard sparseness*. These refer to the distribution of basis vector coefficients for typical inputs. A coding with soft sparseness has a smooth Gaussian-like distribution, but peakier than Gaussian, with many zero values, some small absolute values, fewer larger absolute values, and very few very large absolute values. Thus, many of the basis vectors are active. Hard sparseness, on the other hand, indicates that there are many zero values, *no* or *hardly any* small absolute values, fewer larger absolute values, and very few very large absolute values, and thus few of the basis vectors are active. This is appealing from a metabolic perspective: less energy is used when fewer neurons are firing.

Another measure of coding is whether it is *critically complete* or *overcomplete*. If the number of basis vectors n is equal to the dimensionality k of the input set, the coding is said to be critically complete. In this case, smooth changes in the input vector result in abrupt changes in the coefficients, and the coding is not able to gracefully handle small scalings, small translations, or noise in the inputs. If, however, the number of basis vectors is larger than the dimensionality of the input set, the coding is *overcomplete*. Overcomplete codings smoothly interpolate between input vectors and are robust under input noise. The human primary visual cortex is estimated to be overcomplete by a factor of 500, so that, for example, a 14 x 14 patch of input (a 196-dimensional space) is coded by roughly 100,000 neurons.

Biological Evidence

Sparse coding may be a general strategy of neural systems to augment memory capacity. To adapt to their environments, animals must learn which stimuli are associated with rewards or punishments and distinguish these reinforced stimuli from similar but irrelevant ones. Such task requires implementing stimulus-specific associative memories in which only a few neurons out of a population respond to any given stimulus and each neuron responds to only a few stimuli out of all possible stimuli.

Theoretical work on Sparse distributed memory has suggested that sparse coding increases the capacity of associative memory by reducing overlap between representations. Experimentally, sparse representations of sensory information have been observed in many systems, including vision, audition, touch, and olfaction. However, despite the accumulating evidence for widespread sparse coding and theoretical arguments for its importance, a demonstration that sparse coding improves the stimulus-specificity of associative memory has been lacking until recently.

Some progress has been made in 2014 by Gero Miesenböck's lab at the University of Oxford analyzing Drosophila Olfactory system. In Drosophila, sparse odor coding by the Kenyon cells of the mushroom body is thought to generate a large number of precisely addressable locations for the storage of odor-specific memories. Lin et al. demonstrated that sparseness is controlled by a negative feedback circuit between Kenyon cells and the GABAergic anterior paired lateral (APL) neuron. Systematic activation and blockade of each leg of this feedback circuit show that Kenyon cells activate APL and APL inhibits Kenyon cells. Disrupting the Kenyon cell-APL feedback loop decreases the sparseness of Kenyon cell odor responses, increases inter-odor correlations, and prevents flies from learning to discriminate similar, but not dissimilar, odors. These results suggest that feedback inhibition suppresses Kenyon cell activity to maintain sparse, decorrelated odor coding and thus the odor-specificity of memories.

Neuroinformatics

Neuroinformatics is a research field concerned with the organization of neuroscience data by the application of computational models and analytical tools. These areas of research are important for the integration and analysis of increasingly large-volume, high-dimensional, and fine-grain experimental data. Neuroinformaticians provide computational tools, mathematical models, and create interoperable databases for clinicians and research scientists. Neuroscience is a heterogeneous field, consisting of many and various sub-disciplines (e.g., cognitive Psychology, behavioral neuroscience, and behavioral genetics). In order for our understanding of the brain to continue to deepen, it is necessary that these sub-disciplines are able to share data and findings in a meaningful way; Neuroinformaticians facilitate this.

Neuroinformatics stands at the intersection of neuroscience and information science. Other fields, like genomics, have demonstrated the effectiveness of freely-distributed databases and the application of theoretical and computational models for solving complex problems. In Neuroinformatics, such facilities allow researchers to more easily quantitatively confirm their working theories by computational modeling. Additionally, neuroinformatics fosters collaborative research—an important fact that facilitates the field's interest in studying the multi-level complexity of the brain.

There are three main directions where neuroinformatics has to be applied:

1. the development of tools and databases for management and sharing of neuroscience data at all levels of analysis,

2. the development of tools for analyzing and modeling neuroscience data,

3. the development of computational models of the nervous system and neural processes.

In the recent decade, as vast amounts of diverse data about the brain were gathered by many research groups, the problem was raised of how to integrate the data from thousands of publications in order to enable efficient tools for further research. The biological and neuroscience data are highly interconnected and complex, and by itself, integration represents a great challenge for scientists.

Combining informatics research and brain research provides benefits for both fields of science. On one hand, informatics facilitates brain data processing and data handling, by providing new electronic and software technologies for arranging databases, modeling and communication in brain research. On the other hand, enhanced discoveries in the field of neuroscience will invoke the development of new methods in information technologies (IT).

History

Starting in 1989, the United States National Institute of Mental Health (NIMH), the National Institute of Drug Abuse (NIDA) and the National Science Foundation (NSF) provided the National Academy of Sciences Institute of Medicine with funds to undertake a careful analysis and study of the need to create databases, share neuroscientific data and to examine how the field of information technology could create the tools needed for the increasing volume and modalities of neuroscientific data. The positive recommendations were reported in 1991 ("Mapping The Brain And Its Functions. Integrating Enabling Technologies Into Neuroscience Research." National Academy Press, Washington, D.C. ed. Pechura, C.M., and Martin, J.B.) This positive report enabled NIMH, now directed by Allan Leshner, to create the "Human Brain Project" (HBP), with the first grants awarded in 1993. The HBP was led by Koslow along with cooperative efforts of other NIH Institutes, the NSF, the National Aeronautics and Space Administration and the Department of Energy. The HPG and grant-funding initiative in this area slightly preceded the explosive expansion of the World Wide Web. From 1993 through 2004 this program grew to over 100 million dollars in funded grants.

Next, Koslow pursued the globalization of the HPG and neuroinformatics through the European Union and the Office for Economic Co-operation and Development (OECD), Paris, France. Two particular opportunities occurred in 1996.

- The first was the existence of the US/European Commission Biotechnology Task force co-chaired by Mary Clutter from NSF. Within the mandate of this committee, of which Koslow was a member the United States European Commission Committee on Neuroinformatics was established and co-chaired by Koslow from the United States. This committee resulted in the European Commission initiating support for neuroinformatics in Framework 5 and it has continued to support activities in neuroinformatics research and training.

- A second opportunity for globalization of neuroinformatics occurred when the participating governments of the Mega Science Forum (MSF) of the OECD were asked if they had any new scientific initiatives to bring forward for scientific cooperation around the globe. The White House Office of Science and Technology Policy requested that agencies in the federal government meet at NIH to decide if cooperation were needed that would be of global benefit. The NIH held a series of meetings in which proposals from different agencies were discussed. The proposal recommendation from the U.S. for the MSF was a combination of the NSF and NIH proposals. Jim Edwards of NSF supported databases and data-sharing in the area of biodiversity; Koslow proposed the HPG ? as a model for sharing neuroscientific data, with the new moniker of *neuroinformatics*.

The two related initiates were combined to form the United States proposal on "Biological Informatics". This initiative was supported by the White House Office of Science and Technology Policy and presented at the OECD MSF by Edwards and Koslow. An MSF committee was established on Biological Informatics with two subcommittees: 1. Biodiversity (Chair, James Edwards, NSF), and 2. Neuroinformatics (Chair, Stephen Koslow, NIH). At the end of two years the Neuroinformatics subcommittee of the Biological Working Group issued a report supporting a global neuroinformatics effort.

Koslow, working with the NIH and the White House Office of Science and Technology Policy to establishing a new Neuroinformatics working group to develop specific recommendation to support the more general recommendations of the first report. The Global Science Forum (GSF; renamed from MSF) of the OECD supported this recommendation.

The International Neuroinformatics Coordinating Facility

This committee presented 3 recommendations to the member governments of GSF. These recommendations were:

1. National neuroinformatics programs should be continued or initiated in each country should have a national node to both provide research resources nationally and to serve as the contact for national and international coordination.

2. An International Neuroinformatics Coordinating Facility (INCF) should be established. The INCF will coordinate the implementation of a global neuroinformatics network through integration of national neuroinformatics nodes.

3. A new international funding scheme should be established. This scheme should eliminate national and disciplinary barriers and provide a most efficient approach to global collaborative research and data sharing. In this new scheme, each country will be expected to fund the participating researchers from their country.

The GSF neuroinformatics committee then developed a business plan for the operation, support and establishment of the INCF which was supported and approved by the GSF Science Ministers at its 2004 meeting. In 2006 the INCF was created and its central office established and set into operation at the Karolinska Institute, Stockholm, Sweden under the leadership of Sten Grillner. Sixteen countries (Australia, Canada, China, the Czech Republic, Denmark, Finland, France, Germany, India, Italy, Japan, the Netherlands, Norway, Sweden, Switzerland, the United Kingdom and the United States), and the EU Commission established the legal basis for the INCF and Programme in International Neuroinformatics (PIN). To date, fourteen countries (Czech Republic, Finland, France, Germany, Italy, Japan, Norway, Sweden, Switzerland, and the United States) are members of the INCF. Membership is pending for several other countries.

The goal of the INCF is to coordinate and promote international activities in neuroinformatics. The INCF contributes to the development and maintenance of database and computational infrastructure and support mechanisms for neuroscience applications. The system is expected to provide access to all freely accessible human brain data and resources to the international research community. The more general task of INCF is to provide conditions for developing convenient and flexible applications for neuroscience laboratories in order to improve our knowledge about the human brain and its disorders.

Society for Neuroscience Brain Information Group

On the foundation of all of these activities, Huda Akil, the 2003 President of the Society for Neuroscience (SfN) established the Brain Information Group (BIG) to evaluate the importance of neuroinformatics to neuroscience and specifically to the SfN. Following the report from BIG, SfN also established a neuroinformatics committee.

In 2004, SfN announced the Neuroscience Database Gateway (NDG) as a universal resource for neuroscientists through which almost any neuroscience databases and tools may be reached. The NDG was established with funding from NIDA, NINDS and NIMH. The Neuroscience Database Gateway has transitioned to a new enhanced platform, the Neuroscience Information Framework <http://www.neuinfo.org>. Funded by the NIH Neuroscience BLueprint, the NIF is a dynamic portal providing access to neuroscience-relevant resources (data, tools, materials) from a single search interface. The NIF builds upon the foundation of the NDG, but provides a unique set of tools tailored especially for neuroscientists: a more expansive catalog, the ability to search multiple databases directly from the NIF home page, a custom web index of neuroscience resources, and a neuroscience-focused literature search function.

Collaboration with other Disciplines

Neuroinformatics is formed at the intersections of the following fields:

- neuroscience
- computer science
- biology
- experimental psychology
- medicine
- engineering
- physical sciences
- mathematics
- chemistry

Biology is concerned with molecular data (from genes to cell specific expression); medicine and anatomy with the structure of synapses and systems level anatomy; engineering – electrophysiology (from single channels to scalp surface EEG), brain imaging; computer science – databases, software tools, mathematical sciences – models, chemistry – neurotransmitters, etc. Neuroscience uses all aforementioned experimental and theoretical studies to learn about the brain through its various levels. Medical and biological specialists help to identify the unique cell types, and their elements and anatomical connections. Functions of complex organic molecules and structures, including a myriad of biochemical, molecular, and genetic mechanisms which regulate and control brain function, are determined by specialists in chemistry and cell biology. Brain imaging determines structural and functional information during mental and behavioral activity. Specialists in biophysics and physiology study physical processes within neural cells neuronal networks. The data from these fields of research is analyzed and arranged in databases and neural models in order to integrate various elements into a sophisticated system; this is the point where neuroinformatics meets other disciplines.

Neuroscience provides the following types of data and information on which neuroinformatics operates:

- Molecular and cellular data (ion channel, action potential, genetics, cytology of neurons, protein pathways),

- Data from organs and systems (visual cortex, perception, audition, sensory system, pain, taste, motor system, spinal cord),

- Cognitive data (language, emotion, motor learning, sexual behavior, decision making, social neuroscience),

- Developmental information (neuronal differentiation, cell survival, synaptic formation, motor differentiation, injury and regeneration, axon guidance, growth factors),

- Information about diseases and aging (autonomic nervous system, depression, anxiety, Parkinson's disease, addiction, memory loss),

- Neural engineering data (brain-computer interface), and

- Computational neuroscience data (computational models of various neuronal systems, from membrane currents, proteins to learning and memory).

Neuroinformatics uses databases, the Internet, and visualization in the storage and analysis of the mentioned neuroscience data.

Research Programs and Groups

Neuroscience Information Framework

The Neuroscience Information Framework (NIF) is an initiative of the NIH Blueprint for Neuroscience Research, which was established in 2004 by the National Institutes of Health. Unlike general search engines, NIF provides deeper access to a more focused set of resources that are relevant to neuroscience, search strategies tailored to neuroscience, and access to content that is traditionally "hidden" from web search engines. The NIF is a dynamic inventory of neuroscience databases, annotated and integrated with a unified system of biomedical terminology (i.e. NeuroLex). NIF supports concept-based queries across multiple scales of biological structure and multiple levels of biological function, making it easier to search for and understand the results. NIF will also provide a registry through which resources providers can disclose availability of resources relevant to neuroscience research. NIF is not intended to be a warehouse or repository itself, but a means for disclosing and locating resources elsewhere available via the web.

Genes to Cognition Project

A neuroscience research programme that studies genes, the brain and behaviour in an integrated manner. It is engaged in a large-scale investigation of the function of molecules found at the synapse. This is mainly focused on proteins that interact with the NMDA receptor, a receptor for the neurotransmitter, glutamate, which is required for processes of synaptic plasticity such as long-term potentiation (LTP). Many of the techniques used are high-throughput in nature, and integrating the various data sources, along with guiding the experiments has raised numerous informatics questions. The program is primarily run by Professor Seth Grant at the Wellcome Trust Sanger Institute, but there are many other teams of collaborators across the world.

Neurogenetics: Gene Network

Genenetwork started as component of the NIH Human Brain Project in 1999 with a focus on the genetic analysis of brain structure and function. This international program consists of tightly integrated genome and phenome data sets for human, mouse, and rat that are designed specifically for large-scale systems and network studies relating gene variants to differences in mRNA and protein expression and to differences in CNS structure and behavior. The great majority of data are open access. GeneNetwork has a companion neuroimaging web site—the Mouse Brain Library—that contains high resolution images for thousands of genetically defined strains of mice.

The Blue Brain Project

The Blue Brain Project was founded in May 2005, and uses an 8000 processor Blue Gene/L supercomputer developed by IBM. At the time, this was one of the fastest supercomputers in the world. The project involves:

- Databases: 3D reconstructed model neurons, synapses, synaptic pathways, microcircuit statistics, computer model neurons, virtual neurons.

- Visualization: microcircuit builder and simulation results visualizator, 2D, 3D and immersive visualization systems are being developed.

- Simulation: a simulation environment for large scale simulations of morphologically complex neurons on 8000 processors of IBM's Blue Gene supercomputer.

- Simulations and experiments: iterations between large scale simulations of neocortical microcircuits and experiments in order to verify the computational model and explore predictions.

The mission of the Blue Brain Project is to understand mammalian brain function and dysfunction through detailed simulations. The Blue Brain Project will invite researchers to build their own models of different brain regions in different species and at different levels of detail using Blue Brain Software for simulation on Blue Gene. These models will be deposited in an internet database from which Blue Brain software can extract and connect models together to build brain regions and begin the first whole brain simulations.

The Neuroinformatics Portal Pilot

The project is part of a larger effort to enhance the exchange of neuroscience data, data-analysis tools, and modeling software. The portal is supported from many members of the OECD Working Group on Neuroinformatics. The Portal Pilot is promoted by the German Ministry for Science and Education.

The Neuronal Time Series Analysis (NTSA)

NTSA Workbench is a set of tools, techniques and standards designed to meet the needs of neuroscientists who work with neuronal time series data. The goal of this project is to develop information system that will make the storage, organization, retrieval, analysis and sharing of experimental and simulated neuronal data easier. The ultimate aim is to develop a set of tools, techniques and standards in order to satisfy the needs of neuroscientists who work with neuronal data.

Japan National Neuroinformatics Resource

The Visiome Platform is the Neuroinformatics Search Service that provides access to mathematical models, experimental data, analysis libraries and related resources.

An online portal for neurophysiological data sharing is also available at BrainLiner.jp as part of the MEXT Strategic Research Program for Brain Sciences (SRPBS).

The CARMEN Project

The CARMEN project is a multi-site (11 universities in the United Kingdom) research project aimed at using GRID computing to enable experimental neuroscientists to archive their datasets in a structured database, making them widely accessible for further research, and for modellers and algorithm developers to exploit.

The Cognitive Atlas

The Cognitive Atlas is a project developing a shared knowledge base in cognitive science and neuroscience. This comprises two basic kinds of knowledge: tasks and concepts, providing definitions and properties thereof, and also relationships between them. An important feature of the site is ability to cite literature for assertions (e.g. "The Stroop task measures executive control") and to discuss their validity. It contributes to NeuroLex and the Neuroscience Information Framework, allows programmatic access to the database, and is built around semantic web technologies.

Research Groups

- *The Institute of Neuroinformatics* (INI) was established at the University of Zurich at the end of 1995. The mission of the Institute is to discover the key principles by which brains work and to implement these in artificial systems that interact intelligently with the real world.

- *The THOR Center for Neuroinformatics* was established April 1998 at the Department of Mathematical Modelling, Technical University of Denmark. Besides pursuing independent research goals, the THOR Center hosts a number of related projects concerning neural networks, functional neuroimaging, multimedia signal processing, and biomedical signal processing.

- *Netherlands state program* in neuroinformatics started in the light of the international OECD Global Science Forum which aim is to create a worldwide program in Neuroinformatics.

- Shun-ichi Amari, Laboratory for Mathematical Neuroscience, RIKEN Brain Science Institute Wako, Saitama, Japan. The target of Laboratory for Mathematical Neuroscience is to establish mathematical foundations of brain-style computations toward construction of a new type of information science.

- Gary Egan, Neuroimaging & Neuroinformatics, Howard Florey Institute, University of Melbourne, Melbourne, Australia. Institute scientists utilize brain imaging techniques, such as magnetic resonance imaging, to reveal the organization of brain networks involved in human thought.

- Andreas VM Herz Computational Neuroscience, ITB, Humboldt-University Berlin, Berlin Germany. This group focuses on computational neurobiology, in particular on the dynamics and signal processing capabilities of systems with spiking neurons.

- Nicolas Le Novère, EBI Computational Neurobiology, EMBL-EBI Hinxton, United Kingdom. The main goal of the group is to build realistic models of neuronal function at various levels, from the synapse to the micro-circuit, based on the precise knowledge of molecule functions and interactions (Systems Biology)

- *The Neuroinformatics Group in Bielefeld* has been active in the field of Artificial Neural Networks since 1989. Current research programmes within the group are focused on the improvement of man-machine-interfaces, robot-force-control, eye-tracking experiments, machine vision, virtual reality and distributed systems.

- Hanchuan Peng, Allen Institute for Brain Science, Seattle, USA. This group has focused on using large scale imaging computing and data analysis techniques to reconstruct single neuron models and mapping them in brains of different animals.

- Laboratory of Computational Embodied Neuroscience (LOCEN), Institute of Cognitive Sciences and Technologies, Italian National Research Council (ISTC-CNR), Rome, Italy. This group, founded in 2006 and currently led by Gianluca Baldassarre, has two objectives: (a) understanding the brain mechanisms underlying learning and expression of sensorimotor behaviour, and related motivations and higher-level cognition grounded on it, on the basis of embodied computational models; (b) transferring the acquired knowledge to building innovative controllers for autonomous humanoid robots capable of learning in an open-ended fashion on the basis of intrinsic and extrinsic motivations.

- NUST-SEECS Neuroinformatics Research Lab, Establishment of the Neuro-Informatics Lab at SEECS-NUST has enabled Pakistani researchers and members of the faculty to actively participate in such efforts, thereby becoming an active part of the above-mentioned experimentation, simulation, and visualization processes. The lab collaborates with the leading international institutions to develop highly skilled human resource in the related field. This lab facilitates neuroscientists and computer scientists in Pakistan to conduct their experiments and analysis on the data collected using state of the art research methodologies without investing in establishing the experimental neuroscience facilities. The key goal of this lab is to provide state of the art experimental and simulation facilities, to all beneficiaries including higher education institutes, medical researchers/practitioners, and technology industry.

Technologies and Developments

The main technological tendencies in neuroinformatics are:

1. Application of computer science for building databases, tools, and networks in neuroscience;

2. Analysis and modeling of neuronal systems.

In order to organize and operate with neural data scientists need to use the standard terminology and atlases that precisely describe the brain structures and their relationships.

- Neuron Tracing and Reconstruction is an essential technique to establish digital models of the morphology of neurons. Such morphology is useful for neuron classification and simulation.

- BrainML is a system that provides a standard XML metaformat for exchanging neuroscience data.

- The Biomedical Informatics Research Network (BIRN) is an example of a grid system for neuroscience. BIRN is a geographically distributed virtual community of shared resources offering vast scope of services to advance the diagnosis and treatment of disease. BIRN allows combining databases, interfaces and tools into a single environment.

- Budapest Reference Connectome is a web based 3D visualization tool to browse connections in the human brain. Nodes, and connections are calculated from the MRI datasets of the Human Connectome Project.

- GeneWays is concerned with cellular morphology and circuits. GeneWays is a system for automatically extracting, analyzing, visualizing and integrating molecular pathway data from the research literature. The system focuses on interactions between molecular substances and actions, providing a graphical view on the collected information and allows researchers to review and correct the integrated information.

- Neocortical Microcircuit Database (NMDB). A database of versatile brain's data from cells to complex structures. Researchers are able not only to add data to the database but also to acquire and edit one.

- SenseLab. SenseLab is a long-term effort to build integrated, multidisciplinary models of neurons and neural systems. It was founded in 1993 as part of the original Human Brain Project. A collection of multilevel neuronal databases and tools. SenseLab contains six related databases that support experimental and theoretical research on the membrane properties that mediate information processing in nerve cells, using the olfactory pathway as a model system.

- BrainMaps.org is an interactive high-resolution digital brain atlas using a high-speed database and virtual microscope that is based on over 12 million megapixels of scanned images of several species, including human.

Another approach in the area of the brain mappings is the probabilistic atlases obtained from the real data from different group of people, formed by specific factors, like age, gender, diseased etc. Provides more flexible tools for brain research and allow obtaining more reliable and precise results, which cannot be achieved with the help of traditional brain atlases.

References

- Phil Simon (March 18, 2013). Too Big to Ignore: The Business Case for Big Data. Wiley. p. 89. ISBN 978-1-118-63817-0.

- Russell, Stuart; Norvig, Peter (2003) [1995]. Artificial Intelligence: A Modern Approach (2nd ed.). Prentice Hall. ISBN 978-0137903955.

- Mehryar Mohri, Afshin Rostamizadeh, Ameet Talwalkar (2012) Foundations of Machine Learning, MIT Press ISBN 978-0-262-01825-8.

- Yoshua Bengio (2009). Learning Deep Architectures for AI. Now Publishers Inc. pp. 1–3. ISBN 978-1-60198-294-0.

- Grenander, Ulf. General Pattern Theory : A Mathematical Study of Regular Structures. Oxford University Press. ISBN 9780198536710.

- U. Grenander and M. I. Miller (2007-02-08). Pattern Theory: From Representation to Inference. Oxford: Oxford University Press. ISBN 9780199297061.

- Scherzer, Otmar (2010-11-23). Handbook of Mathematical Methods in Imaging. Springer Science & Business Media. ISBN 9780387929194.

- Peter W. Michor (2008-07-23). Topics in Differential Geometry. American Mathematical Society. ISBN 9780821820032.

- U. Grenander and M. I. Miller (2007-02-08). Pattern Theory: From Representation to Inference. Oxford University Press. ISBN 9780199297061.

- Randall D. Beer (1990). Intelligence as Adaptive Behavior: An experiment in computational neuroethology. Academic Press. ISBN 0-12-084730-2.

- Gerstner, Wulfram; Kistler, Werner M. (2002). Spiking Neuron Models: Single Neurons, Populations, Plasticity. Cambridge University Press. ISBN 978-0-521-89079-3.

- Dupuis, Paul; Grenander, Ulf; Miller, Michael. "Variational Problems on Flows of Diffeomorphisms for Image Matching". ResearchGate. Retrieved 2016-02-20.

- Library, State of Texas, University of Texas Health Science Center at San Antonio. "Albrecht Dürer's Human Proportions « UT Health Science Center Library". library.uthscsa.edu. Retrieved 2016-03-16.

- "White Matter Atlas - Diffusion Tensor Imaging Atlas of the Brain's White Matter Tracts". www.dtiatlas.org. Retrieved 2016-01-26.

Permissions

All chapters in this book are published with permission under the Creative Commons Attribution Share Alike License or equivalent. Every chapter published in this book has been scrutinized by our experts. Their significance has been extensively debated. The topics covered herein carry significant information for a comprehensive understanding. They may even be implemented as practical applications or may be referred to as a beginning point for further studies.

We would like to thank the editorial team for lending their expertise to make the book truly unique. They have played a crucial role in the development of this book. Without their invaluable contributions this book wouldn't have been possible. They have made vital efforts to compile up to date information on the varied aspects of this subject to make this book a valuable addition to the collection of many professionals and students.

This book was conceptualized with the vision of imparting up-to-date and integrated information in this field. To ensure the same, a matchless editorial board was set up. Every individual on the board went through rigorous rounds of assessment to prove their worth. After which they invested a large part of their time researching and compiling the most relevant data for our readers.

The editorial board has been involved in producing this book since its inception. They have spent rigorous hours researching and exploring the diverse topics which have resulted in the successful publishing of this book. They have passed on their knowledge of decades through this book. To expedite this challenging task, the publisher supported the team at every step. A small team of assistant editors was also appointed to further simplify the editing procedure and attain best results for the readers.

Apart from the editorial board, the designing team has also invested a significant amount of their time in understanding the subject and creating the most relevant covers. They scrutinized every image to scout for the most suitable representation of the subject and create an appropriate cover for the book.

The publishing team has been an ardent support to the editorial, designing and production team. Their endless efforts to recruit the best for this project, has resulted in the accomplishment of this book. They are a veteran in the field of academics and their pool of knowledge is as vast as their experience in printing. Their expertise and guidance has proved useful at every step. Their uncompromising quality standards have made this book an exceptional effort. Their encouragement from time to time has been an inspiration for everyone.

The publisher and the editorial board hope that this book will prove to be a valuable piece of knowledge for students, practitioners and scholars across the globe.

Index

www.ingramcontent.com/pod-product-compliance
Lightning Source LLC
Jackson TN
JSHW052158130125
77033JS00004B/189